HORNS, STRINGS, AND HARMONY

ARTHUR H. BENADE is accomplished as a flutist, and as a nuclear physicist, an acoustics expert, and a teacher of science teachers. He is an associate professor of physics at Case Institute of Technology, Cleveland, Ohio.

Born and reared in Lahore, India (now Pakistan), his father was a missionary-physics professor at Punjab University. Dr. Benade's college education started at Ohio State, continued at the University of Minnesota, and included technical work at Los Alamos in 1945 and a six-month mountaintop experiment with cosmic rays near Climax, Colorado. In 1952, he received his Ph.D. degree at Washington University, St. Louis, and joined the physics department of Case Institute of Technology. Dr. Benade now lives in Shaker Heights near Cleveland.

Horns, Strings, and Harmony

Arthur H. Benade

GREENWOOD PRESS, PUBLISHERS
WESTPORT, CONNECTICUT

Library of Congress Cataloging in Publication Data

Benade, Arthur H
    Horns, strings, and harmony.

    Reprint of the ed. published by Anchor Books,
Garden City, N. Y., which was issued as no. S11 of
Science study series.
    Bibliography: p.
    Includes index.
    1. Music—Acoustics and physics. 2. Music—
Physiological aspects. 3. Musical instruments.
I. Title. II. Series: Science study series ; S11.
[ML3805.B33 1979]   781.9'1   78-25707
ISBN 0-313-20771-2

First published in 1960 by Doubleday & Company, Inc.,
New York.
Reprinted with the permission of Doubleday & Company,
Inc.

Reprinted in 1960 by Greenwood Press,
A division of Congressional Information Service, Inc.
88 Post Road West, Westport, Connecticut 06881

Library of Congress catalog card number 78-25707
ISBN 0-313-20771-2

Printed in the United States of America

10 9 8 7 6 5 4 3

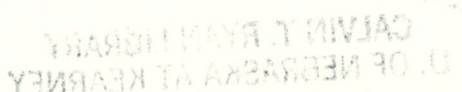

To Virginia,
upon whose taste, counsel, and hard work
I am always dependent

THE SCIENCE STUDY SERIES

The Science Study Series offers to students and to the general public the writing of distinguished authors on the most stirring and fundamental topics of physics, from the smallest known particles to the whole universe. Some of the books tell of the role of physics in the world of man, his technology and civilization. Others are biographical in nature, telling the fascinating stories of the great discoverers and their discoveries. All the authors have been selected both for expertness in the fields they discuss and for ability to communicate their special knowledge and their own views in an interesting way. The primary purpose of these books is to provide a survey of physics within the grasp of the young student and the layman. Many of the books, it is hoped, will encourage the reader to make his own investigations of natural phenomena.

These books are published as part of a fresh approach to the teaching and study of physics. At the Massachusetts Institute of Technology during 1956 a group of physicists, high school teachers, journalists, apparatus designers, film producers, and other specialists organized the Physical Science Study Committee, now operating as a part of Educational Services Incorporated, Watertown, Massachusetts. They pooled their knowledge and experience toward the design and creation of aids to the learning of physics. Initially their effort was supported by the National Science Foundation, which has continued to aid the program. The Ford Foundation, the Fund for the Advancement of Education, and the Alfred P. Sloan Foundation have also given support. The Committee is creating a textbook, an extensive film series, a laboratory guide, especially designed apparatus, and a teacher's source book for a new integrated secondary school physics program which is undergoing continuous evaluation with secondary school teachers.

The Series is guided by a Board of Editors, consisting of Paul F. Brandwein, the Conservation Foundation and Harcourt, Brace and Company; John H. Durston, Educational Services Incorporated; Francis L. Friedman, Massachusetts Institute of Technology; Samuel A. Goudsmit, Brookhaven National Laboratory; Bruce F. Kingsbury, Educational Services Incorporated; Philippe LeCorbeiller, Harvard University; and Gerard Piel, *Scientific American*.

CONTENTS

9

CONTENTS

11

AN AUTOBIOGRAPHICAL PRELUDE

I grew up in Lahore, India (now part of Pakistan), where my parents were Presbyterian missionaries. Because Lahore is a large city with an extremely long history, and because it is the seat of the Punjab University with its many cultural offshoots, it was a wonderful place for a youngster. My contact with science was very early and very natural, since my father was a professor of physics and department chairman at Forman College, an American-supported institution affiliated, in the British manner, with the University. Some of my earliest memories are concerned with "the lab," and with the cosmic ray equipment that was often taken with us during summers in the mountains.

During my early teens a bout with TB kept me out of school for a couple of years, during which I read a grand mixture of literature, trash, history, and science. It was wearisome at the time, of course, but this period proved to be of great value to me later. It helped to give me an appreciation of many things I would otherwise have missed, and prepared me to catch the musical spark lit by a classmate shortly after I was able to return to regular high school classes.

As it did to many people of my generation, the war

cut into and rearranged my college education, which began at Ohio State and continued at the University of Minnesota. The only relevant item in my military career is the year, beginning in March 1945, I spent as a technician at Los Alamos. There were many opportunities to learn physics and other things here, and after Hiroshima there was also the leisure to make a little flute out of plastic.

After the war I completed my undergraduate work in physics at Washington University in St. Louis, where I also met a musically and artistically talented young lady named Virginia Wassall, whom I courted and married. The first semester of my senior year was spent out of class in the mountains at Climax, Colorado, running a cosmic ray experiment on the decay of mu mesons. Up to this time my thoughts had tended to run in the direction of nuclear physics, and I was a little surprised to find myself moving along the well-worn ancestral path! After a start like this, it was only logical to stay on in St. Louis for graduate work, continuing the study of mesons under the direction of Robert Sard.

It was in St. Louis that I first began to think much about the theory of musical instruments, and to try to add a little to our knowledge of them. It was here also that I took on the job of rebuilding a Zumpe piano (built in 1777) which had been broken by an earlier attempt at restoration. Having since had a few students myself, I can now understand the dismay with which this project was received by my thesis adviser!

Shortly after getting my Ph.D. in 1952, we moved to Cleveland. I joined the physics department of Case Institute of Technology, where I am now an associate professor. At Case I have, in the traditional fashion, combined teaching with research in nuclear physics, with occasional bursts of activity in musical physics.

Every summer but the last since coming to Case, I have given a course of lecture demonstrations in modern physics as part of a refresher program for high school physics teachers (sponsored by General Electric). For these lectures I made a great effort to devise as much equipment as possible out of dime-store and scrap-box materials, because it seems worth while to suggest some possibilities to teachers who must work under a small budget. I mention this partly by way of explanation for the spirit of the experiments and for the informal design of the homemade instruments described in this book.

In concluding these introductory notes, I should acknowledge the debt I owe to several musically and scientifically interested teen-agers, in particular Gregory Levin and Stuart Hirsch, whose probing questions and eager curiosity over the past two or three years have helped me to find an elementary approach to the subject of musical physics. I could not have consciously devised a better preparation for writing this book than the sessions I had with these two neighbors. I am also indebted to my colleague Gilbert Krulee for many enlightening discussions on the behavior of the brass instrument family, and to Dr. E. L. Kent of C. G. Conn Ltd. who provided most of the photographs, all the sound spectra, and many of the measurements of wind instruments which appear in this book.

A. H. Benade

Cleveland, Ohio
January 1960

HORNS, STRINGS, AND HARMONY

CHAPTER I

A Brief Look Around

"The secrets of nature betray themselves more readily when tormented by art than when left to their own course."

Roger Bacon (1214?–94)

Preliminaries

When I was a junior in high school, and a beginner at what my teacher called "blowing the flute" (loud, long, and fast!), I discovered Dayton C. Miller's book, *The Science of Musical Sounds*. I had been exposed from my earliest days to the fascinations of science by my physicist father, and my fairly new passion for music was beginning to stimulate me to ask questions about how and why my flute behaved as it did. Professor Miller's book crossed my horizon at a most opportune time. In it I read the names of many explorers of physics who had helped to relate the world of musical artistry to the world of scientific understanding, and I learned that Miller himself was a devoted flute player who had given care to the understanding of his beloved instrument. In the years that have passed, I have kept up my amateur interest in the playing of music, and have developed my interest in physics into a professional one chiefly dedicated to the inner workings of nuclei. But off and on I have given considerable time to thinking about how musical instruments work, and it is quite appropriate that I should now find myself at Case Institute of Technology, in the same physics department over which Miller himself presided for many years before his death in 1941. No doubt his spirit is peering over my shoulder now, as I set down the opening words of a small and informal descendant of his classic work. I should invoke his blessing on my labors, and ask his forgiveness of at least some of my heresies!

In writing this book, I have chosen for myself the role of guide and for you, my readers, that of interested travelers in a strange and colorful land. As a guide, I shall not be able to show you all the towering mountain peaks or quiet valleys in this country, for our time will be limited, but I shall try to point out the roads and warn you against some of the dangers which are to be found. Before we begin our travels, however, I must prepare you with a little talk about the sort of place we are to visit, and also provide you with some equipment for the journey.

Music, like the instruments which make it, belongs mainly to composers and musicians who create patterns of sound for our enjoyment, but there are many other people who can lose themselves in fascination as they try to understand how music comes about, and how it can affect us so profoundly. All these people have their own points of view and their own ways of expressing their thoughts, but their words do not make a chaotic babel since they have collected themselves into fairly recognizable groups, each with its own tribal dialect. Let us see what some of these groups are, before we choose one or another for an extended visit. Among musicians the composer's viewpoint is different from that of a performer, while the man who listens to their joint efforts will have his own way of looking at things. There are also those scientifically minded people who can add the excitement of their professional curiosity to the more sedentary enjoyments of ordinary listeners. Three groups of scientists find themselves thinking particularly about music: the physicists, who snoop into the ways in which objects vibrate to set up sounds in the air; the physiologists, who are interested in the way our ears convert these vibrations into nerve impulses; and the psychologists, who trace these impulses into the brain and try to find out how our minds respond to them.

In our travels we shall be able to visit only the regions inhabited by physicists and by musicians, two groups who sometimes think they should look down on each other. Some people hold that musicians have a subjective and artistic way of looking at things, while the scientific folk are supposed to be detached and objective in their work and speech. Once this has been said, it is easy to belittle one group or the other, depending on your own feelings about the importance of being artistic or objective. Such talk is a wonderful way to start a fight, but it has very little use otherwise, since musicians and scientists are both creators, both judge their work by fairly well-established aesthetic standards, and both are human beings, complete with a well-developed set of foibles. Scientists and mathematicians are constantly talking of "elegant" proofs, and "beautiful" theories. They seek out symmetries and contrasts just as artists do, and they have the same joys of creation. The man who speaks of cold, hard science probably has never done any, knows few scientists, and has read almost nothing about their work from firsthand sources!

Having made this little speech on the unity of science and art, I must now go on to admit that real troubles do arise when physicists and musicians try to talk together. They often use the same words for different things, and they do not always consider the same things important. Now, I do not for a moment want to give you the impression that one or the other of these groups is "wrong" in its ways; on the contrary, they both usually have something to contribute, but we must always make sure we know who is speaking at the moment, and what he means. As is usual in a divided country, there is a sort of "no man's land" between the two parts which we have to cross and recross, and we shall need to look for paths laid out by earlier explorers.

Some Trail Breakers

Perhaps the greatest of these explorers was Hermann von Helmholtz (1821–94), a German military doctor who wrote *The Sensations of Tone,* the foundation book on sound as it is made and heard, and a similar one on light and seeing, carried out some of the most profound mathematical researches into electromagnetic theory, and in addition made a tremendous contribution to our understanding of thermodynamics. After reading all this, you will not be amazed to hear that he was also a talented pianist. A short quotation about this man appears at the beginning of L. S. Lloyd's book, *Music and Sound,* that is particularly appropriate for me to repeat here. It is taken from an address given in 1878 by the great Scottish physicist James Clerk Maxwell, himself one of the most creative of men:

> "Helmholtz, by a series of daring strides, has effected a passage for himself over that untrodden wild between acoustics and music—that Serbonian bog where whole armies of scientific musicians and musical men of science have sunk without filling it up."

Now, I am not in a mood to improve upon the words of my betters, but I may perhaps remind you that Helmholtz had more than daring; he had a thorough, *practicing* knowledge of his subject from several points of view, and he was gifted beyond ordinary mortals. Because most of us are not brought up to recognize classical allusions of the sort educated Englishmen used so freely seventy years ago, I did a little research and learned that Herodotus described the engulfing of armies in the mud of Lake Serbonis, an Egyptian lake which has since dried up. Since many early musical instruments were made of the kind of cane which grows to this day in swampy places around the Mediterranean, we can see that Maxwell chose his words with skill.

Having named a musical scientist first, I must in fairness introduce now a scientific musician, Theobald Boehm. Boehm was trained as a goldsmith in the family business, but very quickly showed his ability on the flute. He made solo concert tours for several years before beginning to feel that many limitations of the flute of his day could be remedied by someone who really worked at the problem. Between about 1830 and 1850 he performed an extensive series of carefully chosen experiments, guided by what little acoustical theory was available to him, and in the end produced an instrument which was essentially the same as the modern flute. Not only does this instrument show the excellence of his researches into sound, but it also reveals Boehm as a first-class engineer of workable and convenient key-machinery. Most of the improvements which took place in the construction of other woodwinds during the last half of the nineteenth century are directly descended from his ideas and from the stimulus of his success. My own first flute was of a pre-Boehm style still manufactured in England until well into the 1930s. I can well remember the feeling of liberation when I changed to a Boehm flute. Let me assure you from firsthand experience that players of flutes, clarinets, and oboes of the older style have to be pretty expert technicians to be able to play smoothly and in tune. The clarinet concertos of Carl Maria von Weber, for example, are real virtuoso pieces when played on the older instruments for which they were written, yet many a modern high-school clarinetist will play them in public on a Boehm clarinet.

There are many other people who have made large contributions to our understanding of music and musical instruments, but I must pass over them silently to make room for others whose work bears more directly on what we shall be seeing. Dayton C. Miller I have mentioned already, without telling you his part in the exploration. During the first three decades of the twentieth century, before there were any of the electronic instruments that make research easier nowadays, Miller studied the quantitative relations between the sounds we hear and the kinds of vibration which set them up. In Chapter III, I shall show you that it is possible to write down a "recipe" for every sort of sound, and we shall also learn how this recipe can help us to understand what is going on. Dayton Miller was the first to get accurate recipes for all the orchestral instruments, and he did it in a way that has earned him a reputation for experimental skill and inexhaustible patience.

Another more or less contemporary man who did a great deal to clarify the ways in which musical instruments work is the French physicist Bouasse. He is not nearly as well known as he deserves to be, although the cause lay partly within his control. He was a peppery, combative sort of man, who never hesitated to say what was on his mind, and he often said it in a way that made enemies. Because of his controversial approach to things he managed to alienate the editors of several journals, and ended up having to publish all his work in book form, printed in small editions and not widely distributed. These books (more than twenty different titles) deal with many branches of physics. About 1929 he wrote a two-volume work entitled *Wind Instruments* and another called *Pipes and Resonators*. These three volumes contain practically everything known at the time about the theory and practice of their subject matter, a great deal of which is his own work or that of his collaborator, Fouché, a skilled musician. It is a real pity that these books are already quite rare, since without them a few people are still going around making meaningless experiments and wrong calculations on things which are clearly discussed by Bouasse. (Bouasse is best known among physicists for his spirited opposition to Einstein's relativity theory. He lost the argument, but he was batting in a big league!)

My list of great explorers has been very short, and does not contain many of the names that are greatly honored by musicians or by scientists. There are two reasons for this: we are not here to make an exhaustive and exhausting survey, and anyhow I prefer to emphasize the sort of men who were able to provide us with a broad view of the territory of musical physics, and who mapped out its rivers and mountains.

You may wonder why it is that I have apparently given more importance to men who explored among the wind instruments than to those who peered into the habits of violins and pianos. It is not just that I am myself addicted to wind instruments; all stringed instruments are built around a single kind of vibrator, the uniform string, whose understanding came first to mathematicians and whose behavior is simple enough that it was described rather early. The beautifully shaped wood casing upon which the strings of a violin are tightened has such complicated behavior, on the other hand, that we should soon get far beyond our depth if we tried to learn about it in detail. The wind instruments have a much wider variety of shapes and vibrations (a wider variety, at any rate, from the point of view of physics), and mathematically inclined explorers have found much more that they could tackle successfully.

Harmony and melody, which are at the heart of music, are like fields and meadows which can be walked in and enjoyed without quite so much toil as is required for mountain climbing, and people as far back as the Greeks have recognized the way in which science and art have grown intertwined. While there are many beautiful flowers to be seen, and stately trees, there is also much marshy ground that is overrun with weedy growths of pseudo-science and of musical faddism. We will see later on that a great deal of the form that music takes is dictated by the way our ears and our nervous system work, in very much the same way that the form of a building is dictated by the law of gravity and the properties of stone. No artist need be ashamed to bow to such dictation. Every art has a form, arising from nature, which requires knowledge and preparation from the artist and from the enjoyer, whether the art is music or painting or physics. This form challenges the creator and guides his efforts, even when he *chooses* to depart from it, and it seems to me very foolish to pretend that an artist is so special a creature that he need not understand the materials of his artistic world.

A Plan for the Expedition

In the remaining chapters of this book I plan first to acquaint you with some of the ideas of physics which we will need in our explorations of musical vibrators; we will then pay a very brief visit to our own ears to learn why they are able to pick out certain kinds of sounds as being different from all the others, and so start ourselves on the road through the elementary principles of harmony. Following this, we will travel quickly through the land of the stringed instruments, and see how it is that the piano is a very special and peculiar instrument, as well as some of the reasons for the versatility of the violin family. Before we press on into the realms of the wind instruments, we will have to stop for some more travel equipment and more maps to show us the way past pipes and horns and reeds, and around some famous mudholes.

Those of you who have already visited music in your scientific reading will find that I will often lead you over relatively untraveled paths. You will, in this way, get a wider appreciation of the terrain, even though a repeat tour might show you more detail. Another and more compelling reason for my different route is my belief that a great deal of future progress in musical physics will come from people who approach it along new paths, taking hints from successful explorers in other, more recently developed branches of physics. My hope is that some of you will find yourselves challenged to take up the exploration on your own, or at least that you will recognize that there is a whole continent waiting to be settled.

Let us now proceed with our preparations for the journey by becoming familiar with some of the properties of the common and simple pendulum, which has been guide and friend to physicists since the days when Galileo was a youth.

CHAPTER II

Simple Vibrating Systems

Most people are aware that vibrating objects can set up disturbances in the air, which then act upon a distant ear so that its owner hears a sound. All musical instruments owe their existence to this fact, and before we can safely approach the musician, we need to provide ourselves with some simple ideas about vibrating objects, about the way in which air is set into vibration, and a little about the ear. We also need to set up a suitable language in which we can speak about the properties and behavior of our musical vibrators. These preliminaries will be taken care of in this chapter and in the following two. After this we can ask ourselves what it is about men and their universe that gives music its general shape, and why certain kinds of vibrators must be chosen if one wishes to make musical sounds.

Pendulums and Hacksaw Blades

Let us begin, then, by looking at a typical and familiar vibrating system, the pendulum, which is simply a weight (called the *bob*) hung on a light rod or a string. When the bob is pulled to one side and released, it swings back and forth in a regular motion, which gradually dies away. Pendulums of different lengths oscillate at different rates, and we must settle upon an orderly way to describe this aspect of their nature. Physicists and engineers usually describe the rapidity of oscillation by giving the number of complete to-and-fro swings (or "cycles") which are made in one unit of time; this number is called the *frequency* of the oscillation. In an ordinary grandfather clock the leftward and rightward swings both give rise to ticks, so that the frequency of ticking is twice the frequency of the pendulum.

If we clamp one end of an ordinary hacksaw blade firmly in a vise, plucking the free end with a finger tip will start it vibrating at a frequency easily followed by the eye, about 5 cps (cps is an abbreviation for "cycles per second"). After an initial twanging sound has died out, the blade moves back and forth smoothly and almost silently. There is no sound to be heard under these conditions for two reasons. First, a hacksaw blade vibrating at a frequency of 5 cps is unheard because the ear does not pass on to the brain any very clear signals indicating the presence of air vibrations having a frequency much below 30 or 40 cycles per second. The other reason is based on mathematical calculations. They show, and experiments verify, that the effectiveness with which vibrating objects can stir up vibrations in the air is very small unless the major dimensions of the object are comparable with the distance which sound can travel in the air during the time of one vibration. The importance of the size of a producer of sound and its relation to the wavelength of the sound it sets up in the air is very much in the mind of a hi-fi enthusiast who needs only a small loudspeaker for reproducing the high-pitched (short wavelength) notes in his record but must have a large diameter woofer for the "lows," which are sounds having long wavelengths. Similar, but usually more complicated, considerations enter the design and construction of musical instruments, where the nature of the "coupling" between the vibrator and the air has a great deal to do with the tone color and even the pitch of the instrument.

Shortening the free length of our hacksaw blade visibly increases the frequency of oscillation, and our ears become more and more aware of a humming sound whose pitch rises as the blade is shortened. When the blade is clamped so that the free end is only 2 inches long, the frequency of vibration is in the neighborhood of 180 cps, and we can plainly hear a sound of definite musical pitch. Here is our first connection between the stimulus applied to the ear and our psychological response to it. The higher the *frequency* of vibration, the higher the *pitch* of the sound seems to us. However, we find that raising the *frequency* of the sound in equal steps (say in increments of 20 cps) would not produce in our ears the sensation of a uniformly rising pitch. The proper explanation of this must wait until later; suffice it to say here that our 180 cps vibration gives a musical pitch lying between F and F sharp below middle C on the piano.

In the $\frac{1}{180}$ second required for a complete to-and-fro oscillation of the saw blade, sound traveling in air at its usual velocity of about 1100 feet per second will have gone a distance of slightly more than six feet; that is, the wavelength of a 180 cps sound is about six feet. The two-inch length of the blade is about $\frac{1}{36}$ of the wavelength of its sound. On the other hand, when the blade was clamped at one end so that it vibrated with a frequency of 5 cps, it was trying to set up disturbances whose wavelength is 1100 feet/second $\times \frac{1}{5}$ second $= 220$ feet. The foot-long blade is then only $\frac{1}{220}$ of the wavelength of its sound, and is therefore a very much poorer source of sound energy. As a matter of fact, it turns out that the shorter, higher-frequency vibrator radiates sound energy into the air about a thousand times more effectively than does the longer blade.

Let us now return to our examination of the pendulum, so as to learn some more of the language we need. The width of swing of a vibrator is often of considerable interest, and our pendulum will serve to illustrate the way in which people customarily describe this.

The maximum distance away from its center, or "equilibrium" position, that the pendulum bob moves during its swing toward either side is called the *amplitude* of oscillation. If, for example, the bob is drawn to one side a distance of two inches and released, the pendulum will be swinging with an initial amplitude of two inches, and as the motion gradually dies away the amplitude of oscillation becomes smaller.

This dying away of the amplitude of oscillation of a vibrator is, of course, an indication that the original store of energy given to it by our hands is gradually being frittered away in friction and also perhaps in the production of sound energy carried to our ears by the air. Let us perform some experiments (real or imaginary) on a pendulum in order to clarify our thought on the way the motion of a pendulum dies.

Suppose that we start a pendulum going with an initial amplitude A; suppose also that we then discover that after ten seconds the amplitude has decreased until it is only $\frac{1}{2} A$. We shall find that after ten more seconds the amplitude will be down to $\frac{1}{4} A$. After 30 seconds the amplitude will be $\frac{1}{8} A$, and after a total of 60 seconds the amplitude will be only $\frac{1}{64} A$. Apparently the pendulum never will come to rest completely, although its amplitude decreases by a half for each ten seconds of time. The real moral to be drawn from this set of observations is that in *equal intervals* of time, the amplitude falls by *equal fractions*. Let us agree to use the words *damping time* for the time during which the amplitude falls to half its original value. A "lightly damped" vibrator—that is, one which hoards its energy closely—will have a long damping time, while a "heavily damped" vibrator rapidly loses its energy to friction and sound radiation, and so has a short damping time.

Some Experiments

In order to make all this new terminology more understandable, let us imagine our pendulum as being mounted in an apparatus like the one in Figure 1A. Suppose a pen is attached to the pendulum bob so that it makes marks on a roll of paper which is slowly pulled past at a steady rate. The wavy line traced along the paper by the pen would then show the position of the pendulum at each instant of time. Figure 1B shows examples of such a recorded line with labels

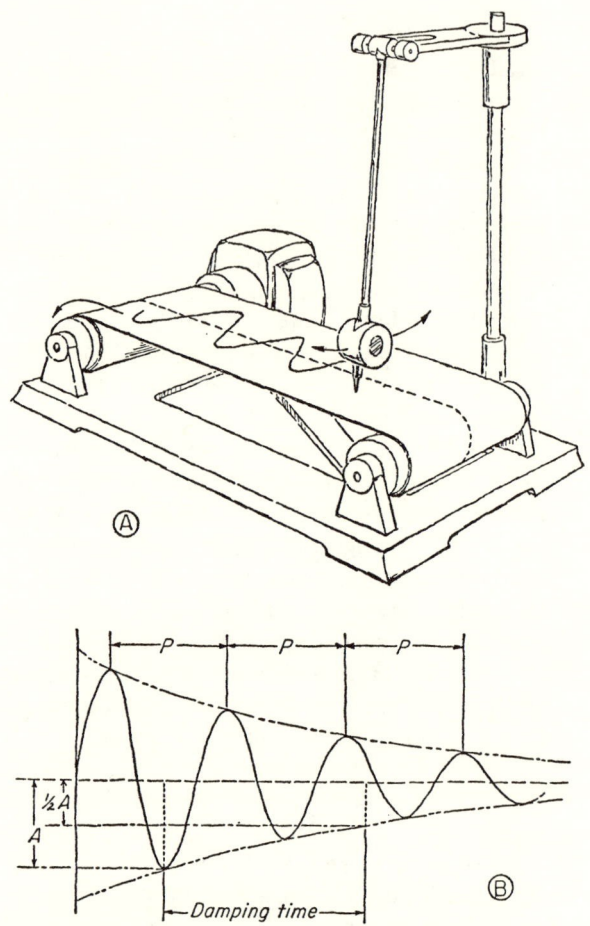

Fig. 1. An apparatus like the device at (A) would trace the motion of a pendulum on a strip of paper moving at a steady rate. A pen is attached to the pendulum swinging perpendicular to the paper strip, which moves uniformly from right to left. The curve shown at (B) is the trace of a damped pendulum. Each equal interval P marks one period (cycle) of the pendulum. Damping time is the time in which amplitude A falls to ½ A.

to show the meanings of the various terms. If the energy losses are negligible, so that each swing has the same amplitude as its predecessor (i.e., the damping time is extremely long), the wavy line represents what is called simple harmonic motion (abbreviated as SHM). It turns out that any conceivable sort of vibratory motion can be built up of a properly chosen set of these simple harmonic motions, each with its own amplitude and frequency.

Because of its importance as a building block, simple harmonic motion is worth more than casual attention, and some more of its properties should be pointed out. For example, the relation of simple harmonic to circular motion is quite striking, as can be seen in experiments like the one sketched in Figure 2.

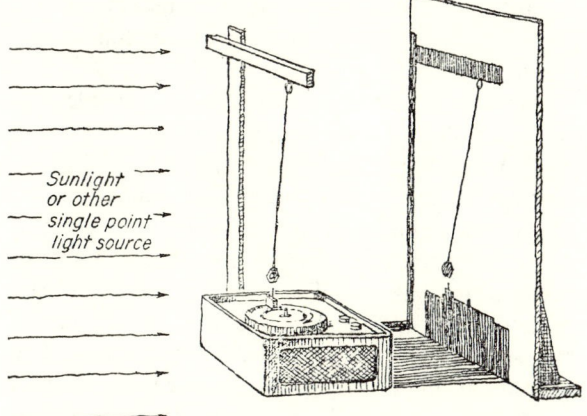

Fig. 2. Shadows cast by the swinging pendulum and by the nail moving with uniform circular motion on the record player turntable show the relation between the two kinds of motion. The shadows on the wall move back and forth together.

Hang a pendulum directly over the center of a record player turntable on which is placed a small object such as a piece of wood with a nail driven up through it, and place a lamp several feet back from the pendulum and turntable so that it casts their shadows on a wall. The length of the pendulum should be accurately adjusted so that it makes a complete oscillation in the time of one turntable revolution (i.e., it should have a frequency of 33⅓ or 45 cycles per minute, depending on the record player). If now the pendulum is pulled aside and released in such a way that the shadow of its bob is directly above the shadow of the revolving nail point, it will be found that the two shadows move back and forth together on the wall. From this we can deduce that the shadow of a circularly moving nail moves in simple harmonic motion with an amplitude equal to the radius of its path, and a frequency equal to the number of revolutions per unit time.

Our experiments have shown us that simple harmonic motion is motion similar to that of the shadow of a circularly moving object. We might also ask whether we can make a general statement about the *forces* that are required to give a motion of this sort to a block of matter. This is, of course, the sort of question that has guided the development of physics since the times of Galileo and of Isaac Newton. These men showed that one can indeed give a concise, and not too difficult, description of the forces required for SHM. A body which feels a leftward restoring force *proportional to its displacement* when it is to the right of its central position, and a similar rightward force when it is to the left, will execute simple harmonic motion if pulled to one side and released. We can clarify our thoughts about the meaning of this by experimenting with a U-tube partly filled with water, as shown in Figure 3. Shaking the tube sets the mass of water into oscillatory motion; let us see that the forces are of the sort required for SHM. At some instant when the water level of the left-hand arm is a distance d below its equilibrium position, we find that there is an unbalanced slug of water with a length $2d$ in the opposite column. The weight of this un-

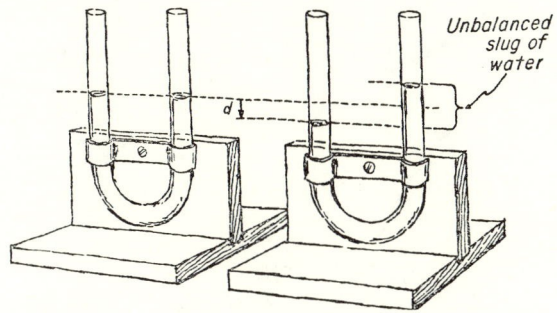

Fig. 3. A U-tube partly filled with water is a useful device for demonstrating simple harmonic motion. When tube is shaken, the mass of water oscillates, the weight of the unbalanced slug forcing the system toward equilibrium.

balanced water is clearly *proportional* to d, and is acting in such a direction as to push the system toward equilibrium. We see from this that a U-tube containing water can oscillate in simple harmonic motion; the moving mass is the whole length of water in the tube, and the proper restoring force is supplied by the difference in level between the two arms. It is a curious fact that the frequency of oscillation of a column of fluid in a U-tube is exactly the same as that of an ordinary pendulum whose length is half that of the length of fluid. This is true regardless of the density of the fluid or of the diameter of the tubing! A thirty-inch length of mercury oscillates at the same frequency as will thirty inches of water whether or not they are placed in tubes of the same diameter, and both will swing in step with a weight hung on a string fifteen inches long.

The next job in founding our musical physics language is to find a way to describe the relation between two simple harmonic oscillations which have the same frequency but are not exactly in step. Figure 4 shows an example of this, where pendulum *b*

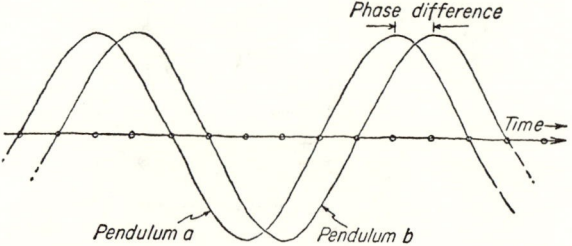

Fig. 4. The curves show the phase difference between two pendulums swinging at the same frequency but out of step. Pendulum b *started its swing a little bit later than pendulum* a *and finished later by the same interval.*

reaches the end of its swing a little bit later than does pendulum *a*. We say that the two pendulums are "out of phase" with each other, the amount being expressed as a fraction of a cycle, or, with the help of our turntable analogy, in degrees. In the example, pendulum *b* is one eighth of a vibration, or forty-five degrees, behind pendulum *a*. In order to produce a pair of shadows moving in simple harmonic motion with a "phase angle" ϕ, we need only place two blocks on a turntable spaced from each other by ϕ degrees on the circle, as shown in Figure 5.

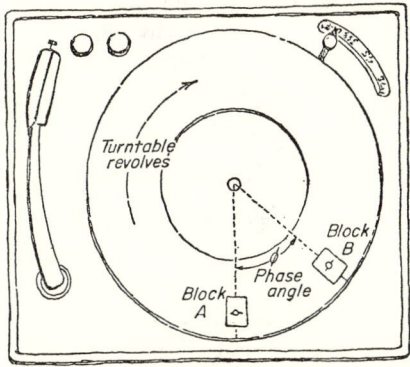

Fig. 5. Difference of phase measured in degrees can be illustrated with two blocks on a turntable. Their shadows on a wall will move in simple harmonic motion with phase angle φ, the angle shown in the illustration.

Connections with Musical Instruments

Musical instruments most closely related to the simple oscillator are those in which some object is set into a vibration that is then allowed to die away. The tuning fork is a prime example of such an instrument. Some other examples are (in order of increasing complexity) the glockenspiel, guitar, and piano. A somewhat less orchestral instrument in this class is the cider jug, which makes a hollow, popping noise when the cork is suddenly pulled out.

Distinct from these instruments are those capable of sustained vibration, the energy of the vibrator being continually replenished by some driving means. The ordinary bowed string instruments, woodwinds, and brasses belong to this latter class. A little study of simple harmonic oscillators driven by external means will prove quite useful when we look into the behavior of these instruments.

Disturbing the Oscillator

By doing or imagining simple experiments of the sort described in the next few paragraphs, we can make an easy acquaintance with the behavior of pendulums when they are disturbed by external forces.

Hang up a pendulum whose "natural" or "free" frequency of oscillation is roughly forty vibrations per minute (the length should be about six and one half inches), two or three feet from a record player arranged as shown in Figure 6. A heavy block of wood resting on the turntable has a nail driven into its top surface in such a way that it revolves in a circle having a diameter between one half and one fourth inch. Tie a string and series of rubber bands to this nail and to the pendulum string about an inch below the point of support; the oscillating force supplied by the turntable will drive the pendulum. The rubber bands are used merely to help you get smoothly varying forces.

Suppose now we run the record player at 33⅓ rpm to drive the pendulum with a force whose frequency is *less* than the free-swinging frequency of the pendulum. After the machinery has been going for a minute or so (during which the initial disturbances can die down), we observe that the pendulum swings *in phase* with the force. It is moving toward the right as the crank pin on the turntable moves toward the right. Any child would of course expect this, since we expect things to move in accordance with the forces we

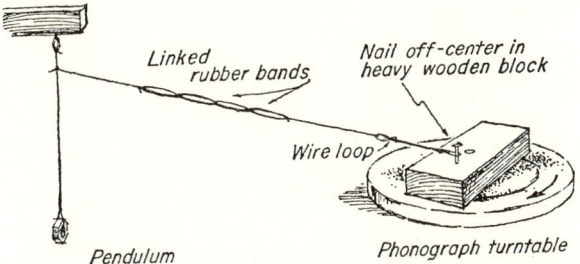

Fig. 6. *Effects of disturbing forces on pendulum swing can be demonstrated with this apparatus. The pendulum should measure 6½ inches from the supporting screw eye to the center of the nut. The string with rubber bands is attached to pendulum string an inch below the screw eye. Use a dozen or so bands and make the string two or three feet overall. The nail on the block should be bent to revolve at a radius of about ¼ inch.*

exert on them. It also seems very natural that the pendulum should swing at the driving frequency (33⅓ cycles per minute) rather than its own natural frequency (about 40 cycles per minute). The important point is that the pendulum swings *in phase* with a driving force whose frequency is *lower* than the free-pendulum frequency.

If, however, the record player is shifted to the 78 rpm setting so that the driving frequency is *greater* than the pendulum's own natural frequency, something new and strange appears. After the usual minute of settling down has elapsed, we find that the pendulum has a motion that is exactly out of step (180 degrees out of phase) with the applied force. The bob appears to come when pushed away, and retreat when pulled toward the turntable! Here we are interested in the fact that the pendulum swings in *opposite* phase from the driving force if the driving frequency is more than slightly greater than the free frequency of the pendulum. These two kinds of phase relation between a vibrator and its driving mechanism will become very important to us when, for example, we examine the behavior of a reed that is attached to some sort of woodwind instrument.

Resonance

It is not very easy to experiment on the amplitude of oscillation set up in the pendulum by forces of different frequency when we use only the apparatus of Figure 6, but if a child's swing is available, we can at least get some idea of what is going on by making a few experiments with it. Suppose someone is sitting quietly on the swing; we have a large pendulum with a certain natural swinging frequency. Now tie a number of rubber bands together end to end in a chain and attach one end to the seat of the swing. Tie a string to the other end of the chain and, standing a few feet away, try to get the swing into oscillation by pulling on the string. If you pull with a succession of rapid tugs, the swing may only jiggle slightly, but will not oscillate back and forth with any appreciable amplitude. A succession of widely spaced tugs has a very similar effect on the swing; however, if the frequency of tugging is arranged to coincide with the natural frequency of the rider in his swing, the swinging will gradually build up in amplitude.

Another and perhaps better way to do the same experiment is for you to sit on the swing and "pump" it by leaning back and forth in the traditional way, at a rate a) faster, and b) slower, than the swing's natural frequency. In neither case will the amplitude build up beyond a slight wiggling at the pumping frequency. A pumping rate that is equal to and properly phased with the natural frequency will, of course, cause the swing to make ever wider oscillations until either your strength or your courage calls a halt. We are dealing here with the phenomenon of *resonance,* the property possessed by all simple harmonic vibrators of responding strongly to regular oscillatory disturbances which have the vibrator's own natural frequency, and of almost ignoring the efforts of oscillatory forces of all other frequencies.

The most familiar application of this general phenomenon is in the common radio receiver. Every radio contains a device in which the current-carrying electrons are able to move in simple harmonic motion with a natural frequency which can be adjusted by the tuning knob. Each broadcasting station sends out an electrical signal at its own definite frequency, and the receiver responds only when it is tuned to one or another of them.

A related effect can be observed if you sing a loud, clear note next to a piano while pressing the loud pedal to raise the dampers. Only one note on the piano will re-echo to your sound to any extent. The string whose natural frequency most closely matches that of the note you have sung will build up in amplitude, and continue to vibrate after the original excitation is shut off. There are certain other notes which also respond weakly, but we must postpone an explanation of this until later.

There are some curious and useful connections between the resonance behavior of a vibrator and its damping time, which can be explained with the help of Figure 7. Here the amplitude of oscillation of our

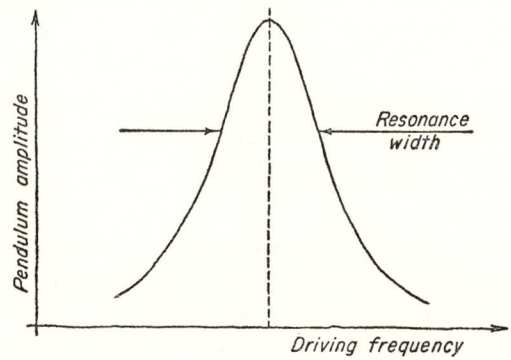

Fig. 7. A vibrator driven at various frequencies gives a resonance curve like this. The amplitude is large only when the driving frequency matches the pendulum's natural frequency. The width of the resonance depends (Fig. 8) on the damping time.

imaginary pendulum is shown as the frequency of the driving machinery is increased. When this driving frequency is low, the pendulum swings to and fro with a small amplitude. As the driving frequency is raised, the pendulum responds more and more strongly, and if the driving frequency is made equal to the natural frequency of the pendulum, it swings with the maximum violence, and has therefore the largest amplitude. The amplitude of the driven pendulum then falls off progressively as the driving frequency is raised above resonance as shown in the curve.

If we have at our disposal a pendulum in which the frictional losses are adjustable, we find that the shape of this "resonance curve" depends on the amount of friction that is present. If the friction is low (so that the damping time is long), we find that the pendulum is very choosy about the driving frequencies to which it responds with vigor. A driving force running only a little above or below the pendulum's own natural frequency excites only a small motion of the pendulum. As if to make up for its selectivity about frequency, a lightly damped pendulum will respond with astonishing violence when driven at its favorite (natural) frequency. On the other hand, a heavily damped pendulum is not nearly so selective about being driven "off resonance," and will respond to driving forces whose frequencies are quite far from the pendulum's natural frequency. However, the amplitude of this damped pendulum is not very great when driven at the resonant frequency. Figure 8 shows these effects, which are observable in all the many sorts of damped oscillators.

The tone quality of many musical instruments is greatly affected by the "widths" of the resonance curves of the musical vibrators concerned, and the ease of playing (bowing or blowing) often depends on the amount of damping present. This is the reason for spending so much time on what might otherwise seem to be a curiosity in the behavior of vibrators.

We have followed the simple harmonic oscillator

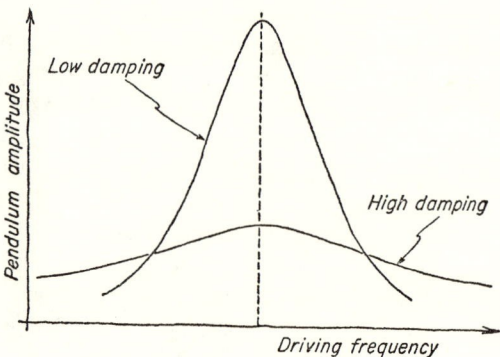

Fig. 8. This graph shows the relation between resonance curve width and damping friction. A vibrator with a long damping time (low damping) gives a large maximum amplitude and a narrow resonance curve. A pendulum system with a short damping time (high damping) has a small maximum amplitude when driven but a large resonance width.

to its lair and found it a relatively harmless creature hiding behind a lot of words. Now we can look forward courageously to more complicated vibrating systems which can move in several ways at once. As usual, physicists keep a set of tame ones for study, to aid them in their hunt for the more subtle vibrators which live in the orchestra.

CHAPTER III

More Complicated Vibrating Systems

Most musical instruments are not simple vibrators of the sort described so far, which have only a single frequency, damping time, and "resonance width." On the other hand, mathematical research into the properties of complicated vibrating systems has discovered ways to simplify them in our thinking. It turns out that the complicated system can, in a certain way, be considered as a set of the simple harmonic oscillators whose properties have already been discussed in so much detail. The mathematics needed to prove this is quite beyond us here; as a matter of fact, even an experienced mathematician will sometimes find it a long and arduous job to disassemble a system into its set of equivalent simple oscillators. But the job can always be done, and, fortunately, the mere possibility often proves to be of great help to us.

Beads on a String

We follow the example of Daniel Bernoulli, a Swiss mathematician, who in 1755 showed a way to sneak up on the problem with the help of some mental experiments on a family of increasingly complex vibrating systems. This family is chosen for the ease with which the separate simple oscillators may be recognized, even though the actual process of proving that they are correct is not simple enough to explain here.

The left-hand column of Figure 9 shows a sequence of systems having increasing complexity. They consist of one, two, or three balls spaced evenly along a taut string that is fixed at both ends. (People who have small children in their homes are at a great advantage in physics, because they can easily steal suitable toys for their experiments. Here, of course, one should get hold of some wooden beads to be placed at two-foot intervals on a string.) If the balls are all drawn to one side to the positions shown in the second column of Figure 9, the balls will feel restoring forces (from the tension in the string) in the directions shown by the arrows. Releasing the balls will allow them to vibrate, and if the amplitude of oscillation of each ball is not too large (less than about 10 per cent of the distance between balls), the vibration will be almost exactly simple harmonic. This vibration, in which all the balls move in phase and with the same frequency will be referred to from now on as being the first, or simplest, or fundamental *mode of vibration.*

The third column of Figure 9 shows a slightly

SYSTEM	TYPE of VIBRATION (complexity)		
Balls on a string	*First mode*	*Second mode*	*Third mode*

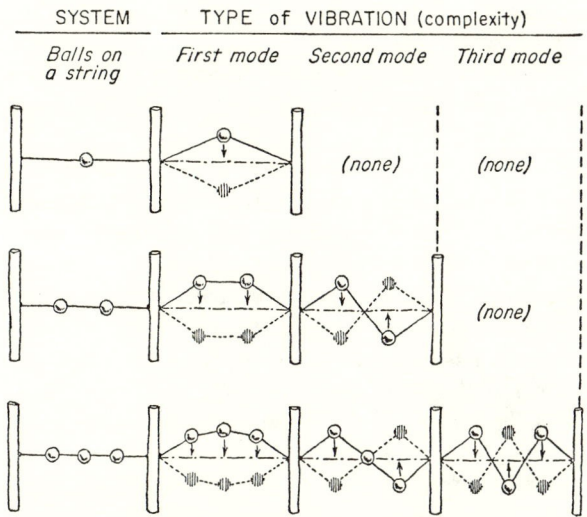

Fig. 9. *Sets of balls on taut strings show different modes of vibration as the system becomes more complex. A single ball finds only one mode of vibration. Two balls can vibrate in two different modes, the second having a frequency 1.732 that of the first. Three balls can vibrate in three distinct ways; the second mode oscillates at 1.847 times the frequency of the first, the third 2.415 times as fast.*

more complicated initial arrangement of the balls before they are released, along with arrows to indicate the direction of the restoring forces which act upon them. Because of its simplicity, the one-ball system cannot vibrate in this more complicated second *mode* of vibration. The frequency of vibration of a given system in this second mode will be somewhat higher than that of the first mode, and the balls will move 180 degrees out of phase with each other in simple harmonic motion of the same frequency. For convenience, we give the label f_1 to the oscillation frequency of the first (or lowest frequency) vibrational mode of a complex system, and let f_2, f_3, etc., stand for the frequencies of the second, third, etc., modes of the same system. The number appearing in the caption for each mode of vibration in Figure 9 gives the vibrational frequency of that particular mode relative to the lowest mode for that system. These are calculated for uniformly spaced balls of equal mass. For example, if the lowest mode of the two-ball system vibrates at a frequency $f_1 = 100$ cps, as determined by the string tension and the mass of the balls, then the second mode of this particular system will oscillate at a frequency $f_2 = 1.732 \times f_1 = 173.2$ cps.

Column four gives an even more complicated initial arrangement of a general sort which is possible only for systems having three or more balls. Again we find the balls oscillating in SHM but at a higher frequency than in the previous two arrangements.

We can deduce several things from this mental experiment with balls on strings, the first being that complex systems have several ways of oscillating in simple harmonic motion. Since we found only one possible mode of vibration for the one-ball system, and three modes for a three-ball system, we can surmise (correctly) that there are always as many modes of vibration as there are balls. In each mode of vibration all the balls move at the same frequency although they may have opposing phase. The higher numbered (more complex) modes of vibration have higher frequencies. Interestingly enough, if we were to increase the frequency f_1 of the lowest mode by increasing the string tension (or by lightening all the balls), the frequencies of all the higher modes would be raised in the same proportion. We shall find later on that this property of weighted strings is one of the phenomena that make pianos and violins possible.

Because the higher modes make more to-and-fro trips per second, they tend to lose their vibrational energy more rapidly than do the lower, simpler modes. In other words, the damping time is generally shorter for the higher frequency modes, the exact amount depending on the kind of friction the system suffers, and on its ability to radiate sound energy into the air. We can make use of this fact in explaining several of the tone-quality peculiarities of musical instruments.

Suppose we now attach a rubber band to the string near one end, and use it to pull the string from side to side with an oscillating force of adjustable frequency. We find the system responding strongly when the driving frequency matches the frequency of one or another of our elementary modes of vibration. Furthermore, we find that the balls actually do move in the manner appropriate to that particular mode, and also that each mode displays a resonance width which is related to its damping time in a way exactly corresponding to the behavior of a simple vibrator.

Several Simultaneous Vibrations

The conclusion we draw from our discussion of our particular system of balls on a string is that *each of its modes of vibration shows all properties of the simple vibrator* which was described in Chapter II. Proper mathematical analysis shows this to be true, and also shows us that *any* conceivable motion of a system of balls on a string can be considered as being made up of a combination of our elementary modes of vibration, which are usually called the *normal modes* of the system. More generally, the behavior of a complex vibrating system can always be described in terms of a properly chosen combination of the normal modes of vibration of that system. For this reason the first job in the analysis of a system is to track down these building-block modes, by fair (mathematical) means or by foul, as we have done in our example.

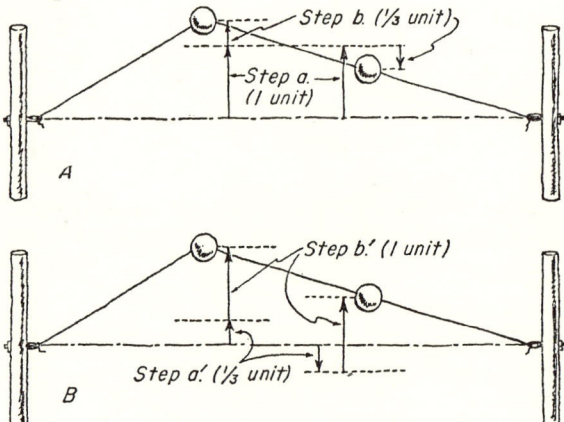

Fig. 10. A particular string shape of a two-ball system can be obtained by combining the initial shapes of the system's two possible modes of vibration. In (A) both balls were pulled up a unit distance; then the left one was raised an additional one third of the distance while the right one was depressed one third. In (B) the left ball first was lifted and the right one depressed one third the unit distance; then both were drawn up the unit distance. The final string shape, as you can see, is the same in both processes.

Let us now use the system of two balls on a string for an illustration of what is meant by the statement that any conceivable vibration of a system can be broken down into a combination of the normal modes of vibration of the system. Suppose (step a) that both balls are first drawn upward a unit distance (say one inch), after which (step b) the left-hand ball is pulled up farther, and the right-hand ball is depressed, both being moved one third of the unit distance, as shown in Figure 10A. This procedure is, of course, exactly equivalent to one shown in Figure 10B in which the left and right balls, respectively, are raised and lowered one third of a unit (step a'), after which they are both raised one unit (step b'). What we have done here is to achieve a certain string shape by the process of combining the initial shapes of the two possible modes of vibration of the two-ball system. (The particular string shape we have set up here happens to be exactly what could have been obtained much more easily by merely pulling the left-hand ball upward a distance of one and one third units, the other ball being allowed to find its own place.) If now the system is released, we find that it will vibrate in the fashion shown in Figure 11, weakly emitting two different sounds, one having the frequency of the lower mode of vibration, one having the higher frequency. (Because of the low frequency of vibration of ordinary sized balls on an ordinary string, these sounds would not of course be audible. We are looking ahead here to more conventional musical vibrators which show similar properties.) The amounts of energy radiated as sound at these two frequencies will obviously depend on the amplitudes of vibration of the corresponding normal modes. The moral to be

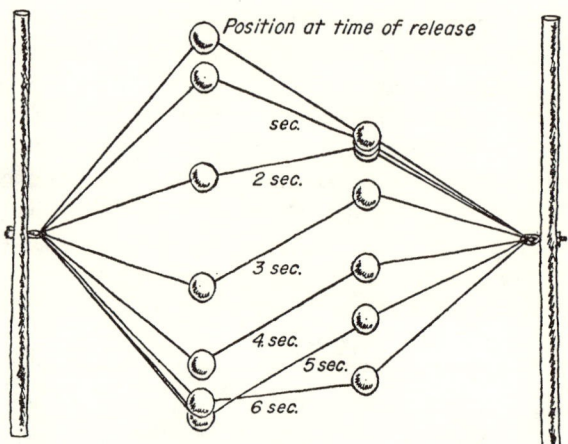

Fig. 11. When released from the configuration shown in Fig. 10, a two-ball system vibrates in this fashion, emitting two different sounds. (The frequency of this system would be too low to be audible.)

drawn from this is that the state in which the system finds itself at the instant it is set into vibration determines how much sound is emitted at each normal mode frequency.

Finding the Recipe

All this can be said more briefly and accurately with some language borrowed from the kitchen. A cake has a certain set of ingredients, and its recipe must contain not only the ingredients needed but also the amounts. Different sorts of cakes can be made using different amounts of the same ingredients. Mathematical physicists have shown that for a given complex vibrating system the complete set of ingredients necessary to make up any of its possible motions is the set of normal modes of vibration belonging to that system. Every imaginable sort of motion for the system is then made by combining different amounts of these normal modes. One can thus write "recipes" for any desired vibration merely by listing the amounts a_1, a_2, a_3, etc., of the first, second, third, etc., normal mode vibrations which are present in the complex vibration. Furthermore, the composite sound produced by the system as it vibrates is made up of a certain definite amount of the sound radiated by each of these normal modes.

Just by looking at a cake or tasting it, a cook usually cannot deduce the list of ingredients with any great confidence, and even less can he guess the amounts of each one present. But a physicist *can* do the equivalent job. He has merely to be patient and clever at finding out what the normal modes of vibration are like, because he knows in advance that these will be the proper ingredients for his vibration recipe. He has also an exact and fairly simple way to find out how much of each ingredient is present. That is, it is not a very difficult problem for him to calculate the set of *a*'s for any vibration that he may observe, once he knows the normal modes of vibration for the system.

Beads and Violin Strings

A vibrating guitar or violin string can perfectly well be thought of as being made up of a very large number of beads strung together, and we can reap the benefits of Daniel Bernoulli's strategy by applying

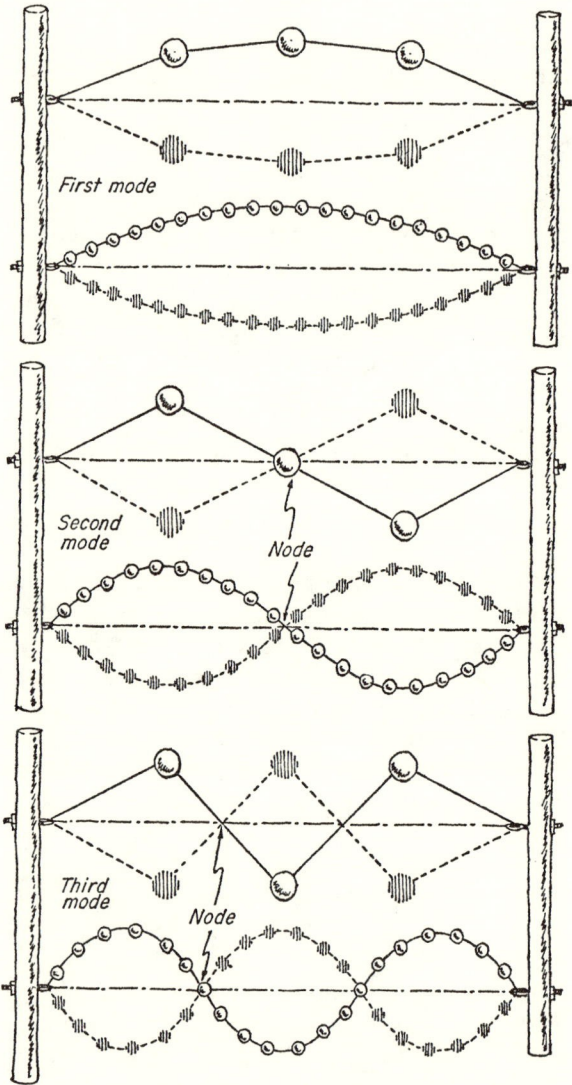

Fig. 12. The three lowest modes of vibration of a string with three balls and a string with twenty balls

our new-found knowledge to these more musical vibrators. Figure 12 shows how closely the lowest three modes of vibration of a twenty-ball string resemble those of the three-ball system we have already examined. It should not take a great leap of the imagination to adapt the pictures to those representing a string carrying a hundred balls, or ten thousand, or on further by easy steps to a smooth string whose molecular (particle-like) structure we can ignore.

I have pointed out that in any given vibrational mode all the particles move in simple harmonic motion at the same frequency. In general, we also find that the different balls move with different amplitudes, and as a matter of fact certain balls never move at all. Figure 12 shows this clearly; for example, in the second vibrational mode the middle ball is always at rest. Whether we speak of multi-particle or of smooth systems, it is often convenient to give a general name to the positions of the stationary portions of the system. These stationary spots are called *nodes,* and one notices that the number of nodes found in the vibrational shape of any particular elementary mode of vibration is always one less than the serial number of that mode. For example, the third mode has $3 - 1 = 2$ nodes, while the thirteenth mode has $13 - 1 = 12$ nodes. Sometimes the positions of maximum vibration in a given mode are called *antinodes* or loops. The serial number of the mode also gives the number of antinodes. This rhyming talk about the relations between modes and their nodes is perhaps confusing,

closely resemble each other. Note that in places the string does not depart from the center line but only wobbles around it. Such places are called nodes.

but, fortunately, we need not worry about it very much.

If the particles are all of the same mass, and evenly spaced along the string, the nodes also are evenly spaced, and the shapes of the various normal modes of vibration for a system with a vast number of balls approach the *sinusoidal* forms we met earlier as the trace of a swinging pendulum. This similarity is not at all an accident, but the connection is only to be found in more advanced mathematical arguments than we have space or background for here. Suffice to say that the swinging pendulum traces a curve that is the graph of the trigonometric function called the *sine;* hence, the particular form of curve is called sinusoidal.

Tapered Violin Strings and Oboes

No ordinary musical instrument uses non-uniform strings that are more massive at one end than at the other, but later on we shall take up various wind instruments (such as the trumpet and the oboe) in which the vibrating air column has a non-uniform cross-sectional area. The general behavior of such air columns is quite similar to that of non-uniform strings, which are much more easily visualized, and so worth some comment here.

Suppose we start out with our old standby, the three-ball string system, but we will change it now so that the balls decrease in mass from left to right, and compare it with a twenty-ball system that has been altered in a similar way. The shapes of the first two modes of vibration of both of these systems are shown in Figure 13. Qualitatively speaking, these vibrational

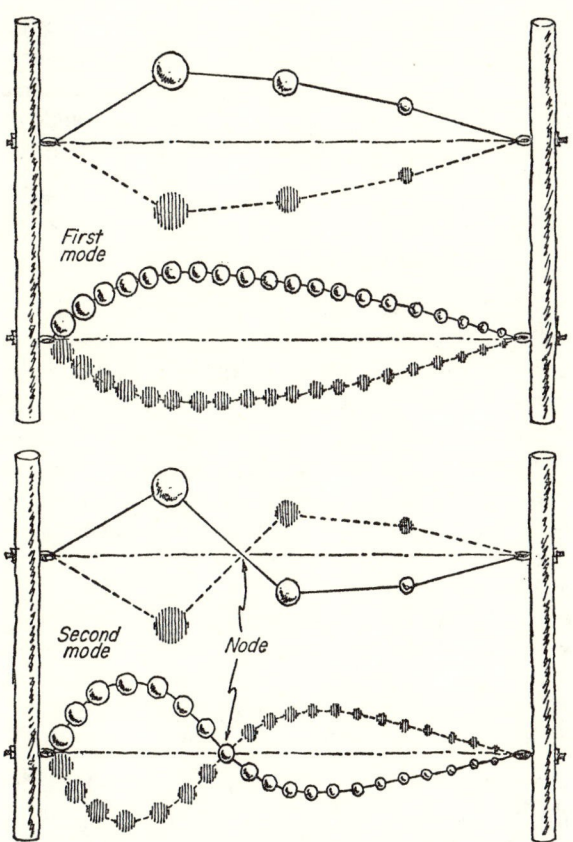

Fig. 13. *Except for a shift of nodes toward the heavily loaded ends of the strings, the vibrational shapes of non-uniform systems are reminiscent of the corresponding shapes of uniform systems.*

shapes are rather like those shown in Figure 12, except that the nodes are all shifted toward the more heavily loaded end of the string. As before, the *nth* mode of vibration has $(n-1)$ nodes, but the vibrational frequency relations between the successive modes do not usually have any particularly simple or interesting properties such as we find for the uniform string.

Back to Musical Strings

When we have studied the musical properties of the ear, we shall learn of an extremely close connection between the usefulness of vibrating uniform strings and the fact that their natural frequencies have a certain simple numerical relation, which we shall now describe. The second vibrational mode of a *uniform* string has a frequency f_2 twice the frequency f_1 of the first mode, the third mode frequency f_3 is three times f_1, and so on. If we call the frequency of the *nth* mode f_n, the frequencies of the normal modes of a uniform string may be summarized by a simple formula:

$$f_n = n \times f_1$$
$$\text{where } n = 1, 2, 3, \text{ etc.}$$

Thus if the first mode has a frequency of 100 cps the fifteenth mode will have a frequency $f_{15} = 15 \times 100 = 1500$ cps, and so on.

Because so many orchestral instruments are built around vibrating strings, we should devote a little space to a description of the "recipes" required to specify the vibration of such a string when it is set in motion in various ways. For example, in a guitar

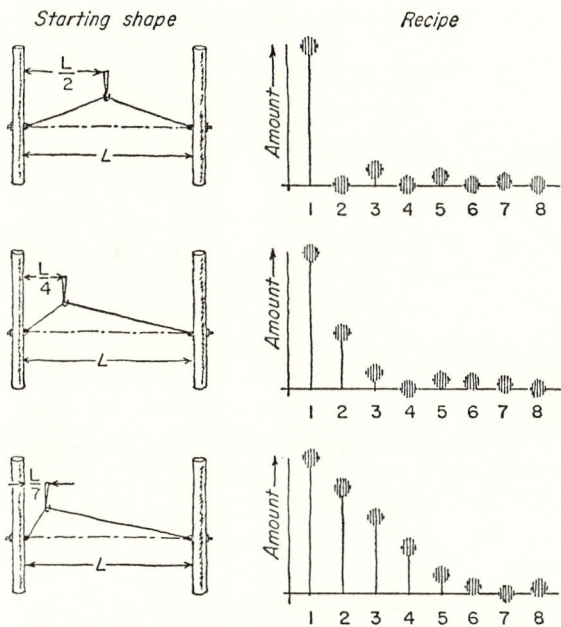

Fig. 14. *Initial string shape determines the vibration recipe. Diagrams at right indicate the amounts of each mode (ingredient) present when the string is plucked in the manner shown at left. When the string is plucked at L/2, modes 2, 4, 6, etc., are not represented in the recipe; similarly, plucking at L/4 gives a recipe without modes 4, 8, 12, etc. This is a general property. Notice that only the first n modes are really important in the recipe for a string plucked at L/n.*

or harpsichord the string is drawn aside at some point and then released. The recipe of the vibration set up in this way and the sound it produces depend on the place of plucking. In Figure 14 are some initial shapes for such a plucked string, together with the corresponding recipes, shown graphically by lines whose heights give the relative amount of each mode needed.

There are, of course, many other ways of setting a string in vibration besides pulling it aside at some given point. You could hold it to one side with a square- or round-ended block before release, or you could strike it, piano-like, with some sort of hammer. Each of these different starting conditions sets up a vibration having a different recipe, which is determined by the shape and position of the block, or the shape, position, and hardness of the hammer. In a real musical instrument the stiffness of the string and some of its other properties alter the initial string shape, and these properties also have an effect on the vibrational recipe.

Anyone who has a guitar or other stringed instrument can hear for himself the variation in sound produced by plucking the different strings at various places with sharp or rounded objects and by striking them with different sorts of hammers. These variations in sound are available to the player of many stringed instruments, to use as he sees fit in the performance of his music. Pianos, harpsichords, and other keyboard instruments do not give the player any appreciable control of the sound of a single played note, but the problem of getting a good sound remains, and must be solved by the instrument maker and his workmen. They use their musical taste to decide what is desirable, and their knowledge of physics to help them achieve it.

The Recipe Changes

So far we have talked about the frequency of each building-block vibration of a given system, and we have seen how the amounts of each ingredient vary with the way in which the system is set in motion. Now we need to figure out how this vibration behaves as time goes on after the system is started. To be precise, we need to notice that the recipe describing the vibration right after starting the system is not the same as the recipe for the motion that is going on a short while later.

Suppose, for example, that a system is started vibrating in such a way that it has equal amounts of the first three normal modes present in its vibration recipe. Figure 15A shows what happens in a simplified way. The curved line at the top of the diagram represents the contribution to the motion that the lowest frequency mode of oscillation makes, and the heavy vertical lines marked a^1, b^1, and c^1 represent the amount of this first mode ingredient in the recipe at the time of starting, after one second, and after two seconds, respectively. The amount present decreases in accordance with the damping time of the first normal mode of vibration. The second curved line represents in similar fashion the contribution made by the second normal mode of oscillation, with a^2, b^2, and c^2 giving the amounts of the second mode in the vibration recipe at the starting time, and one and two seconds later, respectively. As is usual with complex systems, this second mode damps out more quickly than does the first, so that only a little of mode two is left in the recipe after two seconds. The behavior of the third ingredient in the recipe is shown in exactly the same way on line three of the diagram. Because of its short damping time there is very little third mode left after one second, and essentially none present in the vibration recipe after two seconds. Figure 15B summarizes these conclusions with the help of a bar graph, which shows the recipe of the vibration at the three different times.

Musicians, especially the players of instruments with plucked or struck strings, are always concerned with the change of vibration recipe with time. To them it provides an explanation for the really very large change in tone quality which they notice upon striking a note and listening to it as it dies away. In our earlier experiments with plucked hacksaw blades, we had to wait for an initial twanging sound to die down before the blade would oscillate smoothly and quietly. The explanation is now quite simple and obvious: plucking a foot-long hacksaw blade produces a kind of motion with a fairly elaborate vibration recipe. The higher-frequency modes of vibration are easily heard at first because our ears are sensitive to their sounds, and also because the shorter wavelengths of these higher-frequency sounds permit the blade to radiate them efficiently. As we wait, these higher, audible modes of vibration die away, leaving only the 5 cps lowest mode, which is inaudible. A church bell provides a vivid demonstration of the same effect. There is a loud and rather harsh clang when the clapper strikes the bell, the clang being the composite sound produced by the freshly excited vibrational modes of the bell. After a few seconds the higher

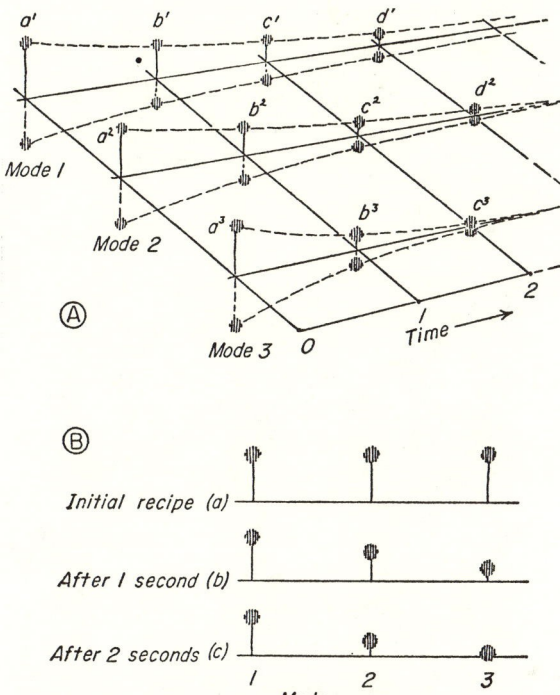

Fig. 15. The vibration recipe can change as time goes on. The diagram at (A) shows the dying away of the first three modes of a system in which the higher modes have a shorter damping time than the lower ones. The diagram at (B) shows the vibration recipes observed at various times after the system is started.

modes have damped out, leaving a soft deep humming, which is produced by the longer-lived lowest mode of vibration.

We have now completed the major part of our preliminary study of vibration theory, or, rather, we have provided ourselves with a set of working ideas and a vocabulary sufficient for a discussion of the physics of music and musical instruments.

CHAPTER IV

Something about Ears

Since music affects human beings with vibrations set up in the air around them, we obviously must pay a little attention to the way our ears are made. Now, ears have many jobs besides the one of providing us with musical enjoyment, and a complete description of the process of hearing would lead us far beyond the fields of physics, through neurology and psychology and on into philosophy. The old principle of "divide and conquer" comes to our rescue here as it does so many places in science. If we temporarily give up the major hope of understanding the whole business of hearing, and divide it into its mechanical, neurological, and psychological parts, we find that we shall do better to take them one by one, and put what we learn together afterwards.

A great deal of what we call music is determined for us by what goes on in the mechanical arrangements of our ears. Let us see then what we can learn by imagining that each ear consists of a mechanical part which responds to vibrations in the air and some-

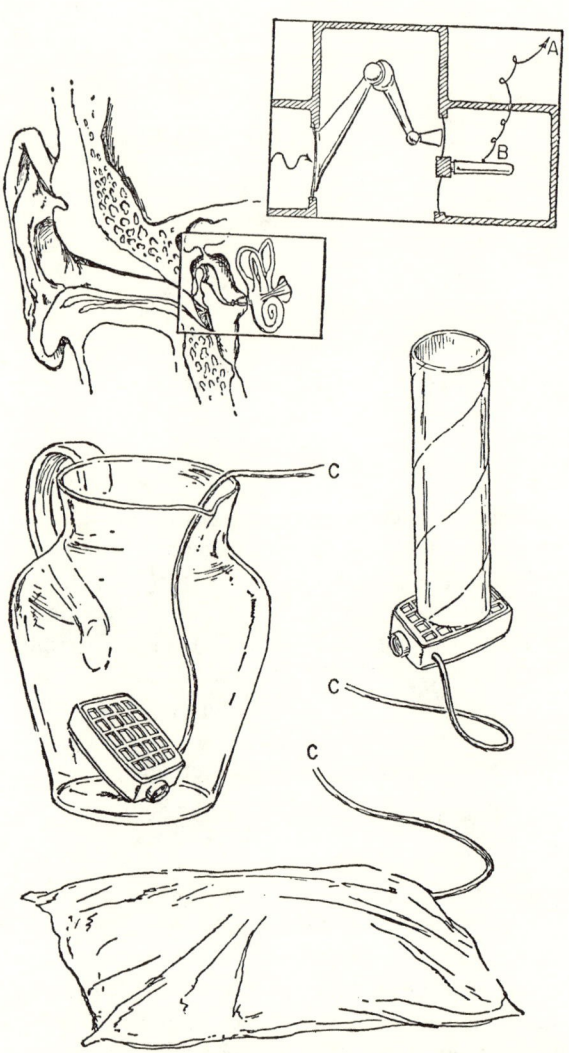

Fig. 16. Simplified mechanical models illustrate roughly the operation and some of the functions of the mechanical parts of the ear. In the diagram at the top the wire ending at (A) stands for the nerve

times alters these vibrations according to the laws of mechanics, together with a part which converts these altered vibrations into nerve impulses going to the brain. If we like, the mechanical part of our ears can be thought of as being a continuation of the outside world. Then we can say that the "true" ear (in some vaguely philosophical sense) is what "hears" the altered sounds coming to it through the mechanical ear. I am quite aware that this stratagem does a great injustice to the many people who have studied the true workings of ears, and my only excuse is that we need to get a quick view of the musical properties of these organs.

Microphones and Ears

Suppose a violinist is playing in a room by himself while we make a tape recording of his music. Obviously, the signals recorded when the microphone is placed on its stand are different from those obtained with the microphone jammed up against one end of a paper mailing tube, dropped into a large pitcher, or wrapped up in a pillow. Here we have several kinds of mechanical connection between musical vibrations set up by the violinist in the air, and the device which converts vibrations into electrical signals for further use. If we want to consider the microphone as being

---

bundle leading from the pick-up (B) *to the brain. In the other diagrams the microphone is the analogue of the part of the ear that converts vibrations into nerve impulses; the glass pitcher, the mailing tube, and the pillow are substitutes for the mechanical part of the ear. Such "attachments" can alter the signal received from a recorder over wire* (C).

analogous to the part of our ears which converts vibrations into nerve impulses, then we can look upon the mailing tube, pitcher, and pillow as being three different substitutes for the mechanical parts of the ear. Figure 16 illustrates these along with an ear. We see at once that what the inner ear gets to hear is only what is passed on to it by the mechanical ear.

Another and perhaps slightly wicked analogy from everyday life is to consider the mechanical part of our ear as a sort of editor of information coming in from the outside world. It deletes certain things and emphasizes others; sometimes it may rearrange this news or even add a few items of its own, in the manner of certain newspaper editors, to slant the messages reaching our minds.

Pitch and Frequency

Before exploring the behavior of the mechanical ear, we should make ourselves understand clearly that our senses respond to physical stimuli, but that our brains do not usually perceive these stimuli in the same way the physicist measures them in his laboratory.

We have seen already that there is some sort of relation between the rate of vibration of an object (a physical property) and the musical pitch (mental interpretation) of the sound we hear. It is a musically useful simplification to say that our minds interpret a sequence of musical tones as being about equally spaced in *pitch* if the successive *frequencies* of their vibrations increase by equal ratios. For example, if the lowest frequency of a sequence is 100 cps, and the next has a frequency of 110 cps, so that the ratio between them is 1.1, then the next tone in the sequence of equal pitch intervals is at a frequency of $1.1 \times 110 = 121$ cps, and the one above that is at $1.1 \times 121 = 133$ cps. The approximate truth of this relation can easily be verified by anyone who has a piano available, because the frequencies of successive notes of the chromatic scale on the piano are carefully adjusted to be in the ratio of 1.05943 to one; each note has a frequency very nearly 6 per cent higher than the note a semitone below it. This type of tuning has been arrived at by perfectly straightforward methods of musical physics, based ultimately on properties of the ear other than the one we are talking about at the moment; so it is fair to use it as a check point here. We find that the pitch change we get by striking first one key and then its immediate neighbor sounds nearly the same to us throughout the middle range of the piano. Toward the ends of the piano keyboard, our simple relationship gets wobbly, and breaks down noticeably, but without causing us any particular trouble during our limited visit to the physical side of music.

Before leaving the subject of pitch, I should mention a phenomenon which causes bickering among musicians in orchestras all over the world. Very loud sounds have a tendency to appear to be at a lower pitch from soft sounds of the same frequency. As a result, a vigorous trumpet player often will accuse his gentler colleagues of playing sharp, because his own notes are so loud in his ears. Every conductor has to mediate this quarrel, and must try to arrange things to satisfy the audience, which hears all the instruments at a much lower level of loudness where the effect is negligible!

In summary, we can say that while the frequency of vibration of the air is most closely connected with the mental sensation of musical pitch, the loudness of the sound may also have an effect.

Amplitude and Loudness

It is not hard for us to believe that vibrators in violent motion (oscillating with large amplitudes) will radiate large amounts of sound energy into the air, and we take it for granted that large amounts of energy reaching our ears should produce sounds which we call "loud." Once again we have to be a little careful to distinguish the physical property (vibrational energy, in this case) from the mental version of it, which we call loudness.

Let us try, or imagine, an experiment that is an exceedingly crude version of one first performed in 1837 by the physicist E. H. Weber, and much more recently by Professor S. S. Stevens of Harvard University. Assemble in one room several radio receivers all tuned to the same station, and mark the volume controls at the settings that make them all sound equally loud *when played separately*. Listen to one of them alone, and then turn up the volume control of another one to the mark. There will be a certain fairly small increase in loudness, which corresponds to the addition of the energy from the second radio. If now a third receiver is turned up to its marked volume setting, we find that the loudness does not appear to increase as much as before, even though there is exactly the same increase of sound energy. Figure 17 is a diagram showing the numbers of radios required to produce equal increments of loudness. This experiment, at first glance, would horrify any competent acoustician. But if all the radios are of different make and are placed randomly in a fairly hard-walled room, the experiment does indeed take on a semblance of truth. Obviously the loudness of a sound depends on the amplitude of the disturbances reaching our ears, but once again, as with frequency and

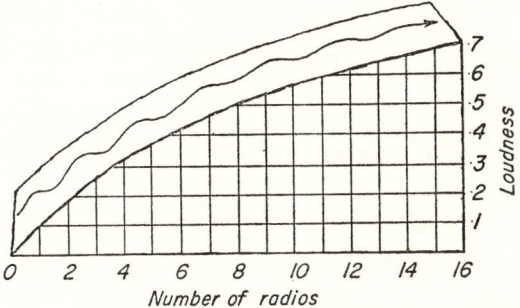

Fig. 17. The apparent loudness of several radios played together does not increase evenly with the number of radios, as this graph indicates. It takes eight radios to sound five times louder than a single radio.

pitch, our brains have their own ways of classifying sounds. These ways are related to but not identical with the methods engineers and physicists use.

While the amplitude of a sound vibration is the chief determinant of the sensation of loudness, the frequency of the vibration also gets into the act and complicates things to an extent which itself depends on loudness! For example, the vibrational energy required for a softly played note whose frequency is near 130 cps (about an octave below middle C on the piano) needs to be about 100 times as great as the energy required to produce the same loudness in a note whose frequency is near 2100 cps (the third C above middle C). On the other hand, the energy ratio required for equal loudness of these two notes when both are loudly played decreases to almost unity. The loudness dependence of the "frequency response" of our ears is a thorn in the side of the hi-fi lover, since it requires him (if he is a purist) to readjust the tone controls of his equipment when he changes the volume at which he wishes to hear his records.

Vibration Recipes and the Ear

We are all able to recognize the voices of our friends. Some of us can foretell from warning sounds that our cars are about to break down, and, of course, music lovers take great pride in being able to tell each other whether it is an oboe or an English horn which is being played at a given moment in a concert. We learn to do these things quite well regardless of the loudness of the sound, and, generally speaking, regardless of the pitch. Several of the giant steps Helmholtz took in his journey from physics to music and back were based on Georg Ohm's discovery that what we have called the vibration recipe provides the explanation of our ability to recognize sounds as being different even when they appear to have the same pitch and loudness. (Ohm was a nineteenth-century German physicist best known for his "law" concerning electrical conduction.) Each of our friends speaks with the help of a complex vibrating system, which gives rise to a recognizable vibration recipe composed of certain amounts of a set of building-block vibrations. Similarly an engine produces a sound whose vibration recipe is determined by moving parts which roll, slide, and knock against one another in a regular pattern in time to the revolutions of the crankshaft. Each musical instrument vibrates in its own complex fashion giving rise to its especial recipe, which was developed by a process of survival of the fittest to play some part in the world's music.

Once again we see that a physical property of the sound reaching our inner ears, the vibration recipe, has its mental counterpart: the "tone quality," "tone color," or "timbre" of the sound. Before we turn to the editorial behavior of our mechanical ears, I should like to summarize the leading facts about the relation of our mental sensations to the various physical stimuli which call them forth.

PHYSICAL PROPERTY   MENTAL PROPERTY

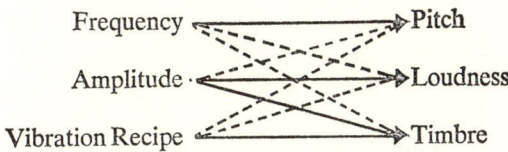

Frequency → Pitch
Amplitude → Loudness
Vibration Recipe → Timbre

Sounds from Two Simple Vibrators

Music's charm and interest come, not from the successive sounding of simple oscillators, but rather from combinations of sound produced by sets of these elementary vibrators which give the characteristic tone colors of the various instruments, and from the arrangement of notes from these instruments into melodies and harmonies. In other words, we take pleasure in patterns of sound built up, in the manner of a mosaic, by musicians, out of the elementary vibrations so dear to the hearts of physicists. All this makes a music-loving physicist itch, musically and scientifically, until he can get to work on the physics involved. Artists are sometimes made uncomfortable by remarks like this, so I should perhaps return a moment to the mosaics I mentioned. The Byzantine craftsman who was so skillful in arranging little shards of pottery or bits of glass and stone into pictures probably thought he was free to follow his soul without let or hindrance, unconscious of the fact that even he was concerned with such gross and materialistic matters as the optical reflectivity and the brittleness of his materials, the geometrical laws of perspective (which he often honored in the breach, but yet with effect), and the physiology of the human eye. The case is very similar for music and musicians, except that here the relations are perhaps closer and more binding, as we shall see in the next chapter.

We must find some simple-minded yet suitably instructive situation on which to practice our curiosity. Let us start therefore by finding out what the ear does when two simple harmonic vibrators are sending sounds in its direction. Two tuning forks will provide an ideal pair of vibrations, although gently blown soda-pop bottles, whistles, or flutes will do reasonably well, since their vibration recipes contain essentially only one single ingredient. Suppose that, at a certain instant, the vibrating part of our ears is being pushed backward and forward in step with the air pressures generated by the two forks, both of which are assumed to be vibrating at exactly the same frequency and amplitude. The ear mechanism will, under these conditions, be driven to oscillate in a simple harmonic motion whose amplitude is essentially double that which would be set up by the sound from either fork acting alone. If, however, one fork vibrates with a slightly lower frequency than the other, the slower fork will gradually fall farther and farther behind its cousin, until it is momentarily a half cycle behind. At this time the ear feels a force in one direction from one fork, and an equal force in the opposite direction from the other, so that there is no vibration at all set up in the ear and no sound is heard. The fork of lower frequency continues to fall behind until it has fallen a complete cycle behind, and the initial condition of double-amplitude vibration is repeated. In such an experiment we hear definite alternating swells and lulls of the sound. This phenomenon usually goes by the name of "beating," and the number of alternations of loudness occurring per second is called, logically enough, the *beat frequency*. A little counting on the fingers, or an examination of Figure 18, convinces us that the beat frequency is always equal to the *difference* in the frequencies of the two sound sources.

Since there is an apparently related phenomenon that gives what are called "difference frequencies," let us try to stall off the traditional confusion by redescribing beats. When two simple oscillators of equal amplitude and *nearly equal* frequency are sounded together, we are aware of a *single* sound having definite pitch whose frequency is the average of the frequencies of the two original sounds, and whose *loudness* grows and shrinks with a frequency equal to their difference. THERE IS NO PART OF THE SYSTEM THAT IS VIBRATING AT THE BEAT FREQUENCY, so the vibra-

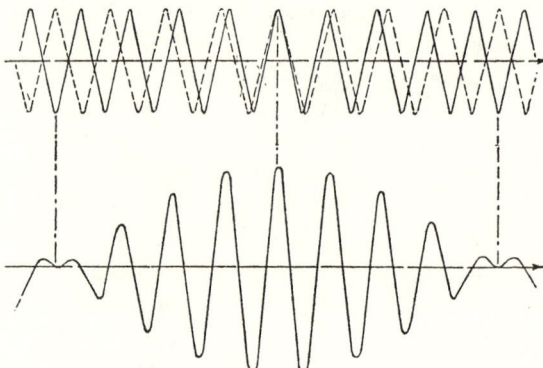

Fig. 18. The beat phenomenon produces these curves. In the upper one, two simple harmonic forces of slightly different frequencies get in and out of step. In the lower graph the resultant of the two forces appears to grow and diminish at the beat frequency.

tional recipe of the composite sound does not contain a component at the beat frequency.

If you want to hear beats, go to the nearest piano and strike a single note; the three strings belonging to this note were tuned quite accurately to have the same frequency (by adjusting the beat frequency to zero) at the last visit of the piano tuner, but a few days later things have probably shifted enough to give slow beats —one or two per second. The typical jangly sound of a single note on a barroom piano comes from similar beats which are too rapid to be followed individually by the ear.

Let us pause now and recognize that we have passed the first milestone within musicians' territory. We have discovered a physical reason for the simplest of all musical intervals, the unison between simple harmonic sounds! You may sneer, and say it is a trivial matter, but we must begin somewhere, and this is our port of entry. We have found that two simple harmonic vibrations, softly played, sound rough and jangly when fairly close to each other in frequency, giving rise to countable beats when they are within a few cycles of each other, and blending into a smooth hum when they have exactly the same frequency. We have found that there is something very special about the physical behavior of the ear, and about our mental response, when a unison is played. We shall find that there are many other such special responses to certain pitch intervals in music, and that many of them come about ultimately through the phenomenon of beats. These special-sounding relations between notes are the focal points of harmony, since they stand out like beacons against the undistinguished darkness of randomly combined tones.

I could write a whole chapter about the way in which man and nature seem to be most interested in, and affected by, the special, the unusual, and the particular. A hawk sees its camouflaged prey on the ground only when it moves. Light always travels through lenses along the path which requires the *least* time of all conceivable paths. Cooled water vapor cannot condense in the middle of the empty air, but must wait around for a speck of dust or an ion to make one point in space special and distinct, and therefore a fitting place to start a raindrop. Music could be said to have condensed around the special pitch intervals.

The Ear Creates

At the beginning of this chapter I remarked that sometimes the mechanical ear makes up sounds of its own to pass along to our brains. We need to look into this "creative" action of our ears, because several of the pillars of music are based upon this property.

Suppose we listen to a source of very pure simple harmonic sound of fixed frequency as its loudness gradually increases. At first we hear only the soft smooth humming sound that is characteristic of the tuning fork, a rather uninteresting and colorless sound of very little musical use. When the loudness is up to the level of a vigorously blown bottle, we begin to notice that the tone color of our sound has changed, becoming a little brighter and more piercing. Some people with musically trained ears will say that they can hear an admixture of the "octave" in the sound, and as the loudness increases they may also hear the "twelfth." In physicists' language this translates to the statement that the vibration recipe for the sound *entering our brains* now contains two new frequencies which are double and triple that of the external sound source. These new frequencies are generated in the mechanical ear by virtue of the fact that the tissues of the ear do not move an amount proportional to the pressure which acts on them. The ear is said to be "non-linear" in its response to sounds. This means that a simple harmonic sound wave sets up ear vibrations which are not simple harmonic, and so require several ingredients in their recipe. We find by calculation and by experiment that when the ear is subjected to a single, loud, simple harmonic sound of frequency f (say 200 cps) it will produce within it-

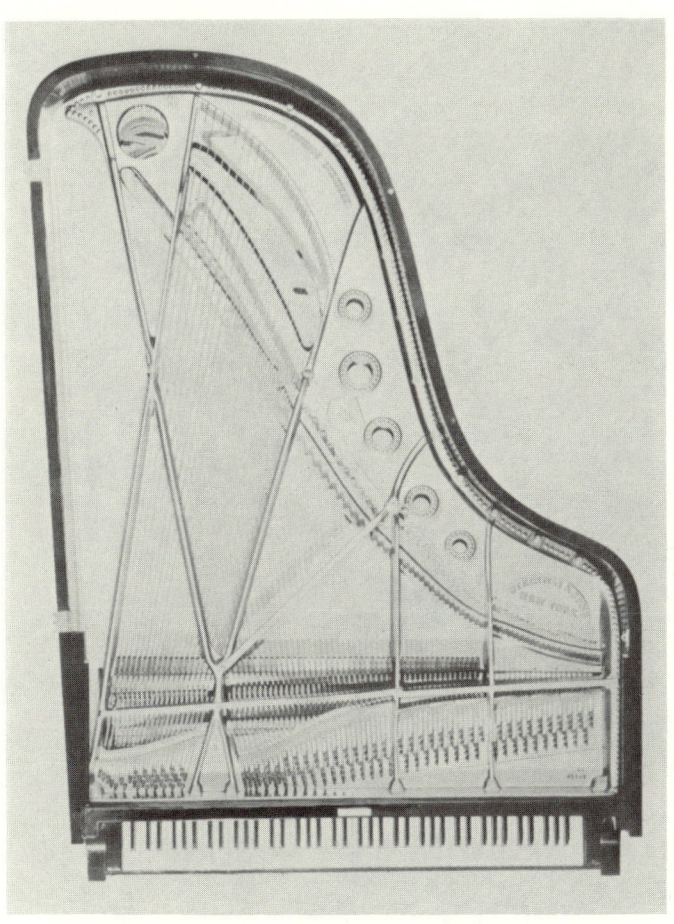

Plate I. The string distribution, bridge placement, and frame bracing against string tension are clearly shown in this photograph of a grand piano's inner mechanism. (Courtesy of Steinway & Sons)

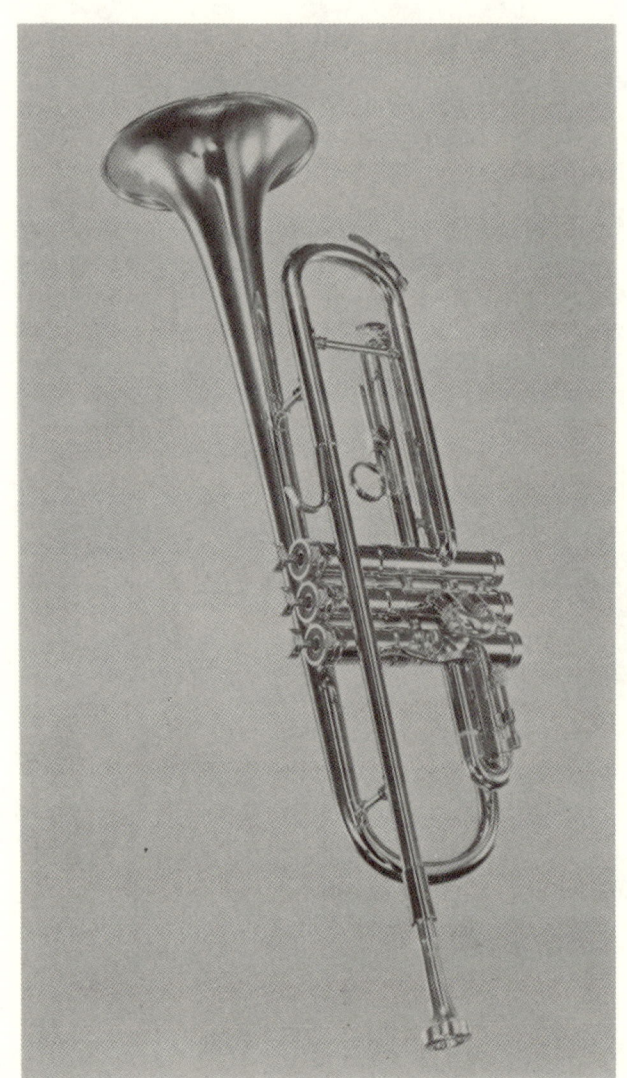

Plate II. Trumpet (Courtesy C. G. Conn Ltd.)

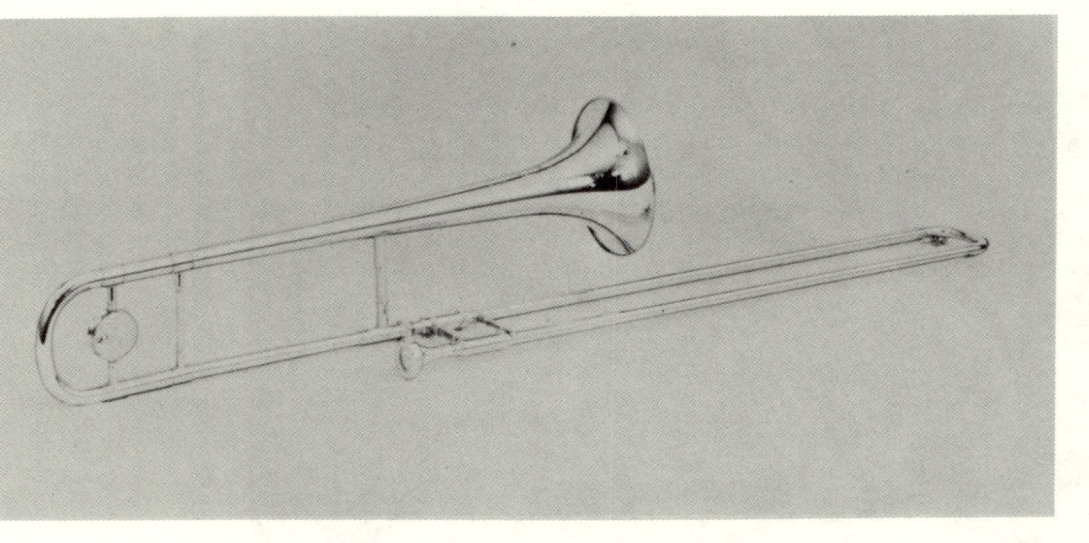

Plate III. Trombone (Courtesy C. G. Conn Ltd.)

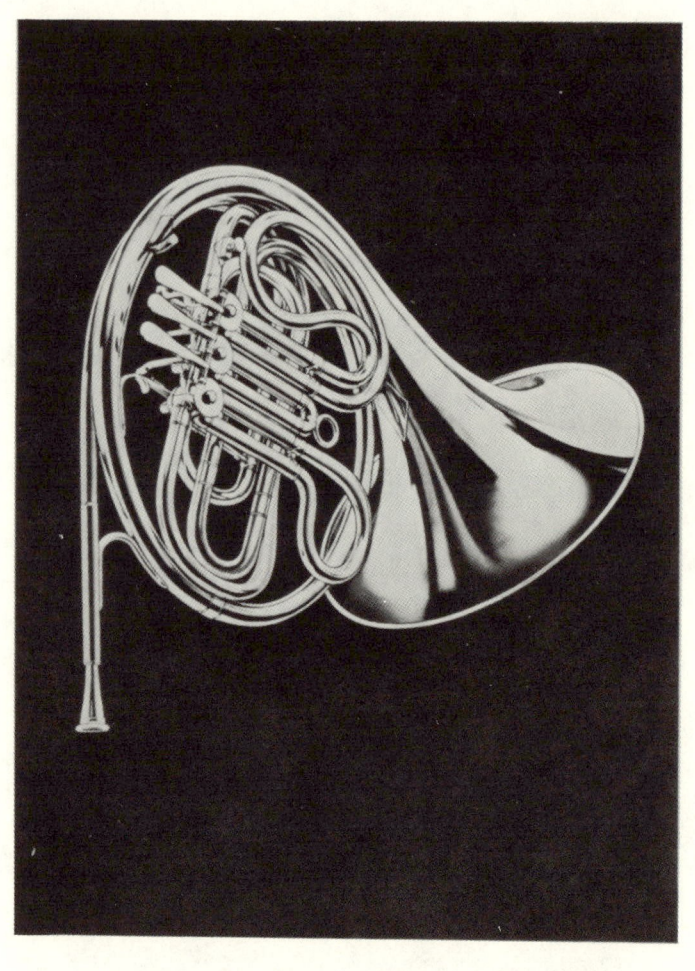

Plate IV. French horn (Courtesy C. G. Conn Ltd.)

self a set of new frequencies belonging to the family $2 \times f, 3 \times f, 4 \times f,$ (400, 600, 800 cps) etc. If the pitch of the loud sound supplied to our ears is that of piano middle C, the new notes generated by the ear's mechanism will be:

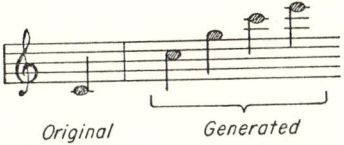

Original   *Generated*

It is not an accident at all, as we shall see, that such a sequence of notes is exactly the one that bugles and other brass instruments are designed to produce.

If two loudly played simple harmonic sounds enter the ear, one of the frequency f and the other of frequency F, we hear not only f and F, but also members of the more extensive family set forth in the table below.

TABLE OF COMBINATION FREQUENCIES

f, 2f, 3f, 4f, etc. F, 2F, 3F, 4F, etc.	"purebred" descendants of f and F
$f - F, f - 2F, f - 3F$, etc. $F - f, F - 2f, F - 3f$, etc. $f + F, f + 2F, f + 3F$, etc. $F + f, F + 2f, F + 3f$. etc.	"crossbred" descendants of f and F
2f − F, 2f − 2F, 2f − 3F, etc. 2F − f, 2F − 2f, 2F − 3f, etc. 2f + F, 2f + 2F, 2f + 3F, etc. 2F + f, 2F + 2f, 2F + 3f, etc.	more complicated descendants of f and F

For musical purposes we really need keep track only of the parts of this chart enclosed by dotted lines, because only these frequencies are usually loud enough to matter.

Let us remember that all these new "combination" frequencies actually exist as vibrations somewhere in the ear, even though older books sometimes referred to them as "subjective" tones because their origin was not clearly understood.

Let me describe a simple little experiment that will help us later in seeing the musical importance of our new-found and unruly family of combination frequencies. If I play a loud pure tone of frequency f together with a very soft one of adjustable frequency F, I notice that whenever the "exploring tone" frequency F is set near to a multiple of f (that is, when F is nearly equal to $2f$, $3f$, etc.), I can hear an ordinary beat occurring between F and the nearby multiple of f! This in itself is a direct proof of the real existence of the new frequencies, and, as a matter of fact, all the important members of the combination-tone family for *two* loud sounds shown in the table can be sought out and located by a third, softly played, "search tone."

The most easily heard of all the combination frequencies is the so-called "difference frequency" $f - F$, the one which people often mix up with the beat frequency. We recall, however, that the beat is produced between two gentle notes which are very close together in pitch, and we hear a *single* note of intermediate pitch that swells and shrinks in loudness at the best frequency. But here, with loud sounds coming in, we can distinguish not only the two separate incoming sounds, but also a new note at the difference frequency. The two incoming sounds can be quite far apart in pitch and still produce a "difference note."

An easy way to get a difference note is to tune a pair of pop bottles (by filling one with water) in such a way that blowing on them produces a pair of notes differing from each other by an amount about equal to the pitch interval between C and E on the piano. When they are blown loudly together, you will hear a low-pitched buzzing sound in addition to the bottles' own notes. This new buzzing sound will be pitched about an octave below the sound from the bottles.

A swanee whistle (slide whistle) can be used with a fixed pitch whistle to show all sorts of difference tone effects. Start with the two whistles loudly played and far apart in pitch, and slide the one toward the other in pitch. The difference tone will start high and swoop downward until it becomes too low-pitched to hear. A somewhat frivolous example of the musical use of this behavior is that given here:

The top line gives the opening notes of *Yankee Doodle,* with the second line a monotonous sequence of C's. If these two lines are played together loudly on tonettes or flutes (any pair of woodwinds will show the effect quite clearly), your ear will produce a third "part" as accompaniment, of the sort shown approximately in the third line of notes.

A more serious example by a first-rate composer (who did not consciously know of the combination-

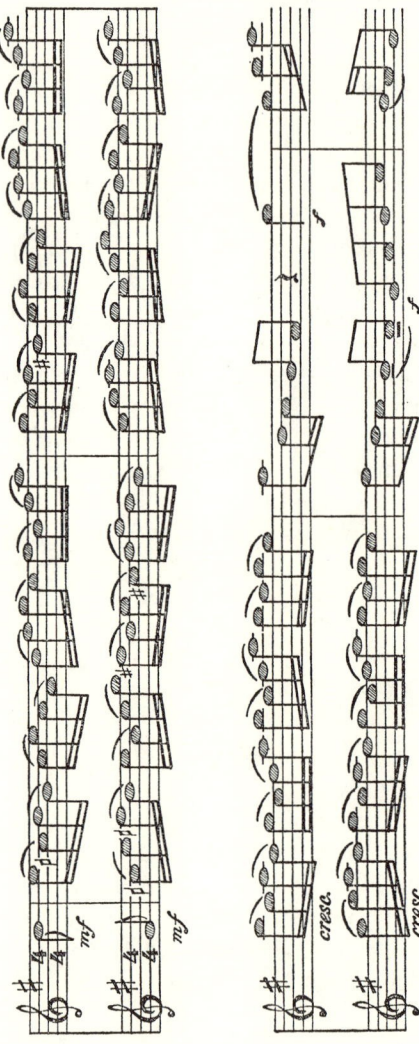

Georg Philipp Telemann: Sonata IV in E Minor

tone effect) is the facing extract from one of Telemann's sonatas.[1]

This rather abrupt chapter ending signals our arrival at the edge of music, and the finish of our study of the physics needed to travel in reasonable safety through the land of Euterpe.

[1] Georg Philipp Telemann, "Six Sonatas for Two Violins, Flutes or Recorders," Sonata IV in E Minor, *Affettuoso,* ed. Josef Marx, pub. Weaner-Levant, New York, 1944.

CHAPTER V

Ears: Architects of Harmony

If we listen to melody and harmony all over the world, and through history, we find so many things that turn up over and over that we are entitled to suspect the existence of physical causes for them. The Greek philosopher Pythagoras was one of the earliest we know of who thought about this in an orderly way. He based his theory of music on the fact that some of the musical intervals (that is, *differences* in pitch) between notes in the Greek scales seemed to be related in a direct way to the sounds given out by a string when it is plucked with some simple fraction of its length left free to vibrate. In more modern language, we say that a half-length string sounds the important interval of one octave in pitch above the full-length string, and a string which is constrained to vibrate over only one third of its length sounds higher by the musical interval of a "twelfth," and so on, making what musicians call a harmonic series of notes. We shall return to the meaning of this a little later on.

Suffice it to say that a certain amount of number juggling can show that almost all the musically important intervals of modern western music (and of most other forms as well) can be more or less deduced from the whole-number fractions of the lengths of a vibrating string.[1] Now, it must be confessed that in order to do this one used to need not only a great deal of faith in the magic properties of numbers, but also a good knowledge of the musical customs which were to be explained. To the Greeks, and to many people who came later, the magic of pure number was so potent that they were quite willing to settle for the numerical relations they found, without ever asking *why* they came out that way. Some musicians with more logical correctness than physical knowledge have seized upon the Pythagoreans' need to know the musical answer beforehand as evidence that the numerical method of organizing music is a sort of peculiar accident. They are tempted to assert that all musical rules are a matter of custom and agreement, and that there is no absolute right or wrong way to go about choosing notes in music. My own belief is that music is *very* strongly affected by the manner in which things vibrate, and by the manner in which our ears work. It is certainly easy enough to show mechanical reasons why certain pitch intervals are special and distinct from their neighbors, in a musical sense, as I propose to do in this chapter.

Before going forward with this task, I should, however, remind you that I shall make use only of some

[1] This is very clearly explained in Chapter IX of Alexander Wood's elegant little book, *The Physical Basis of Music,* Cambridge University Press, 1913.

mechanical facts of life.[2] I shall say nothing about all the complex things which go on between the mechanical ear and the brain. These also have a big effect on music, but would lead us out of our depth and away from the physics of our subject.[3] Even so, we shall find many clear indications that musicians and composers remain members of the physical universe even when their music is "out of this world"!

Special Intervals between Successive Simple Tones

Let us start by asking what happens when we play successively two simple harmonic sounds of differing frequency. If the first note is played quite loud, we hear not only the frequency of that note, but also integral multiples of that frequency, $2f$, $3f$, $4f$, and so on, as explained in the last chapter. You will recall that the "non-linear" response of the ear mechanism is responsible for this behavior, and that the higher multiples are relatively weak in comparison with the lower ones. If now the second note played happens to be the same frequency as *one* of the family of sounds produced by the first note, we recognize it (unconsciously) as a repetition of something heard before, and so consider a note of this second frequency as being in a special relation to the first one.

Everyone who has fooled with a guitar, or played

[2] One of the major contributions of Hermann von Helmholtz to musical physics was a systematic and quantitative analysis of the relations we are about to explore. These are clearly set forth in great detail in his monumental book, *The Sensations of Tone.*

[3] A lively book which goes into these matters is the recently published *Waves and the Ear,* by Willem A. Van Bergeijk, John R. Pierce, and Edward E. David, Jr. (Science Study Series, 1959).

a violin, knows that the pitch rises as the vibrating length of a string is decreased, and many of you will have heard in school that the frequency of vibration of a string is inversely proportional to the length of the string. This means that cutting down the length to one half will double the frequency, reducing it to one third will triple the frequency, and so on. We can now see very clearly why Pythagoras' experiment with strings might be expected to correlate with his knowledge that there are certain especially interesting musical pitch relations.

You may feel that this little demonstration is not enough to lay the scoffing ghosts of untrammeled artistic freedom, and you are right. But it is a step in the proper direction, and a bigger step than it may appear at first. I am being a little rude here toward people who make a hobby of not knowing enough about their subject of specialization. Composers who, for example, preach the extremist dogma of the "twelve-tone scale," in which every note is on an even footing with every other one, are no better in their disregard of the physical world than are sculptors in wood who try to ignore effects of the grain on the texture, working, and strength of their carvings. These musicians have, of course, already seriously undercut their position by admitting the existence of octaves! We shall, however, find a couple of slightly extenuating circumstances in Chapter VI.

Notice that in our experiment with two simple harmonic sounds we only assumed that the first note of the pair was played loud enough to produce its family of "harmonics," as they are called. The second note could be of *any loudness at all,* although we would find new relationships coming in weakly if both notes were loud. In general all we really need is that the loud note be the lower one in frequency. Due to the decreasing loudness of the higher-frequency members of the generated recipe, the importance of the special relations decreases as the two original sounds are separated more and more in frequency.

Two Simple Tones Played Together

Let us now ask what happens if two simple harmonic sounds are played *together*. As before, we shall assume that one is quite loud, and the other one is rather softly played. The loud one will once more be accompanied by its set of generated harmonics, and if the soft one has a frequency very near one member of this set of frequencies, we hear a beating sound at the frequency of that member, in the way described in the last chapter. It is obvious that we get a rather special tonal effect from the two played notes if their frequencies are related in a way that puts the soft note exactly in unison with a note from the loud sound's retinue, since the beating then will cease. The relations we find in this way are, of course, identical with those which we found in the last section, but they are much easier to demonstrate in home experiments since we can directly observe the disappearance of the beats.

Once again slide whistles, tonettes, recorders, or flutes, because of their nearly simple vibration recipes, are best for experimentation. Play two of these at the same pitch, one loudly and one softly, and notice that they beat if the tuning is not exact. Then slide up in pitch with the softly played whistle until it is again in a condition of "zero-beat," which will be found one octave higher than the fixed, loudly played instrument.

90

A region of noticeable beating surrounding a point of smooth, beatless sound will discriminate the next one of the special pitch intervals; it will be found at a pitch about a half octave higher yet (triple the frequency of the loud, simple sound). In musical notation we could summarize this by saying that if the two whistles start out on C, raising the pitch of the softer whistle will show us special effects at the next C above, followed by others at G, C, E, etc. The higher members of this sequence will not be very prominent if they show at all, unless the whistle is blown exceedingly hard.

Some Complex Sounds Are Pleasant

I am sure that you are beginning to get tired of reading about simple harmonic tones, and perhaps also of listening to them in artificial little experiments. We all know perfectly well that it is difficult to find sources of simple, single-ingredient sounds in the everyday world, and it is even harder in the musical world. Musicians generally dislike and avoid such sounds—an orchestra made up of tuning forks, marimbas, flutes, and triangles would give out horribly monotonous music. Each musical instrument has been chosen over the years to produce a certain complex sound, with its own characteristic vibration recipe, depending also on the particular instrument, on its player, and on the effect he wishes to obtain.

But why should a violin be acceptable and a squealing brake unacceptable as a source of musical sound? They work in exactly the same way, with a rubbing device that feeds energy into a complex vibrator. We are led to ask about the nature of the vibration recipes of the two devices, and to inquire what the ear will do with the sounds it gets from each. When we do this, and look around in the musical world for confirmation, we find that the ear can detect a certain kind of physical order in almost all the sounds that are considered musical; on the other hand, unmusical sounds are almost universally of a type that the ear does not find orderly in this sense. Interestingly enough, it also turns out that the orderly properties the ear can find in a single musical-sounding note are precisely the same ones that make for well-defined harmonies and definite scales.

For thousands of years instruments have used stretched uniform strings, and we are quite safe in taking the vibration recipe produced by such a string as an example of musical sound, and in trying to discover its distinctive physical features. In Chapter III I told you that the normal modes of vibration of a uniform string were such that the second mode oscillates at twice the frequency of the first (lowest) mode, the third mode is at three times the lowest possible frequency, and so on. The frequencies of the various modes, in other words, are integral multiples of the lowest mode frequency. Furthermore, I said that any conceivable vibration of the string could be cooked up with the help of a vibration recipe composed solely of the normal modes of vibration. It is a *property* of our *ears* that makes these uniform strings useful in music. I can show why certain other sorts of vibrators are not so useful.

Suppose that we have chosen a string that vibrates at a frequency of 100 cps when oscillating in its lowest mode, and let us give this frequency the name A for future reference. The name of the second mode will be *B,* with a frequency of 200 cps; we can also label the next few modes *C, D,* and *E,* these having frequencies of 300, 400, and 500 cps, respectively. If we strike the string, it will vibrate with all these modes present to some extent in its recipe, and our ears are exposed to a simultaneous set of simple har-

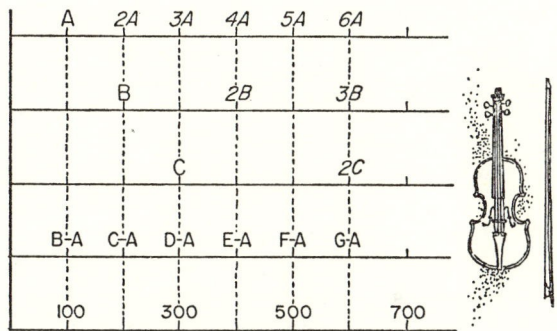

Ingredient frequencies

Fig. 19. The frequencies created by the ear from the sound of a vibrating uniform string are related in this chart to the string's vibrational frequencies. The first three lines show the multiple-frequency components generated from the string harmonics A, B, and C. The fourth line shows some of the difference frequencies generated. The bottom line shows that the resulting ingredient frequencies are identical with the string's own vibrational frequencies.

monic sounds having the frequencies *A, B, C,* etc. A loudly played string will then give our ears the excuse to do a little creative work of their own, which we can study with profit.

We need to refer back to the chart of combination frequencies that appeared on page 81; with its help we find that our ears will take the frequency A and use it to generate new frequencies, $2A$, $3A$, $4A$, etc., as shown in the first line of Figure 19, along with the numerical values of these frequencies. A similar thing will happen to the simple harmonic sound of frequency B which comes from the string, as is shown in the second line of Figure 19, and the third line shows the similar behavior of the ear toward C. Not only does the ear produce these new sounds, but it also generates some with frequencies $B-A$, $C-A$, $D-A$, etc., which are shown in the fourth line of the figure. There are of course many other combination frequencies, but they all behave in the same way. Notice that every sound signal that goes on to the brain has a frequency which is a member of the original set of integral multiples of A, even though its amplitude may be altered. No matter what the ear does to the sound of the vibrating string, the recipe is *still made up of exactly the same ingredients!* Another way to look at it is to notice that every sound component entering the brain is related to every other in one of the simple special ways which we found important when listening to pairs of simple sounds. The net result in the brain is somewhat like a hillbilly family reunion; the crowd gets along very well because everyone is simultaneously brother, cousin, uncle, and grandpaw to everyone else!

Figure 20 shows in a similar way the first few combination frequencies for a vibrating system for which the normal mode (ingredient) frequencies are *not* related by whole numbers; the ear makes a real hash of the vibration recipe it sends on to the brain. Not only do the different vibrations sent in by the source refuse to recognize any musical kinship, but their combination-note descendants are mean and ill-favored, bearing no musically decipherable resemblance to their parents. The result of playing such a sound is confusion, further confounded by what seems almost like malice when it is played loudly.

Let us summarize what we have found, and as usual extend our special case to a more general truth. To a considerable degree, the pleasantness of a sound

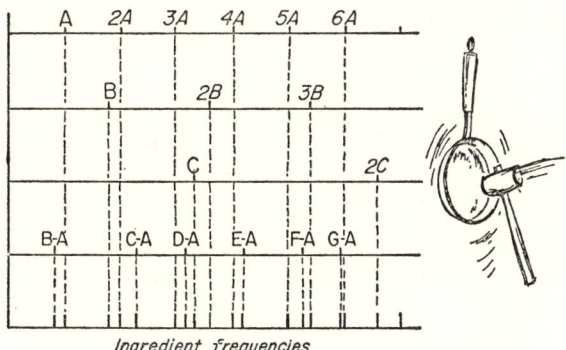

Ingredient frequencies

Fig. 20. The frequencies that the ear creates from the sound of an "unmusical" vibrator are charted here. Notice that the ingredient frequencies that our ears generate are not at all the same as the vibrator's own natural frequencies.

and its usefulness in music are governed by the relations between the frequencies in its vibration recipe. If these frequencies are related to one another by the ratios of integers, the sound is harmonious. The ear recognizes the relations between the components, and will also prove able to appreciate the relations of one such complex sound to another when several instruments are played together. On the other hand, most common objects vibrate with recipes having none of the simple arithmetical relations which Pythagoras was so eager to find. These "disorderly" sound generators are not admitted to the orchestra, except to shock people once in a while. From this I may draw another moral. I have spoken of the majority of complex vibrators as being disorderly, where I mean that the ear does not find simplicity in the recipe it generates for itself, and yet the vibrator is faithfully following the orderly path laid out for it by the laws of dynamics. Order and disorder are relative words, and depend on the context. A child might arrange a set of books according to size, and be very pleased with the effect, but the same effect might leave a librarian tearing her hair. The point is that the order that interests us is the one that has a relevance to our purposes at the moment.

Because *musical* sounds are, as we have seen, all of a kind with regard to their vibration recipes, we find it useful to invent some words to express the special properties of these sounds. Let us call the lowest frequency sound present in a musical tone the "fundamental frequency," or simply the "fundamental" of that tone. The other frequencies present (double, triple the fundamental frequency) we shall call the "harmonics" of the fundamental. Thus a frequency three times as large as the fundamental will be called the third harmonic; one seven times as large will be the seventh harmonic of the fundamental in question. (It is an interesting fact that our minds almost invariably consider the *pitch* of a complex musical note as being the pitch of the fundamental. This is so even if the fundamental is missing from the vibration recipe. Perhaps you can figure out why; look at the difference frequencies!) Always be careful to remember that the *properly defined* harmonics are *exactly* integral multiples of the fundamental. Our new language applies only to a rather narrow class of vibrating systems. Since this is an important bit of terminology, let me explain it all over again. A complex system can vibrate in a whole set of different ways, called modes of vibration. We generally rank these in order of increasing vibrational frequency, and the frequency of the lowest mode is often called the "fundamental" frequency. If the system is such that its higher modes vibrate at frequencies which are integral multiples of the fundamental, we can say that these higher frequencies are *harmonics* of the fundamental. As an example of a system whose higher modes are harmonic, we have met the uniform string. A string loaded with three balls does *not* have its higher modes harmonic, as we saw in Chapter III (see Fig. 9).

Musical Sounds Played Together

Music, as I have said several times, is made up of collections of sounds which have certain relations between them of such a sort that the ear can recognize and appreciate them. Western music in particular (western as distinct from oriental, not western in the sense of cowboy) has developed an elaborate system for combining sets of simultaneously played notes into chords. It is on this basis that we find composers writing symphonies and other large works with many instruments playing many different notes, and it is necessary to adhere rather closely to the rules, if the music is not to sound "wrong" to the listener. Of course, a composer is entitled to put in some of these "wrong" notes for effect (there is a very famous group of despairing, "wrong" sounds from the horns in the second movement of Beethoven's third symphony), but the effect is of the same sort as that of pepper in cooking. A little gives interest and zest, but pepper is hard to live on as a steady diet.

There are no doubt hundreds of intelligent and thoughtful people who would happily drown me for these superficial remarks on the rules of music, and I freely admit that there is some good music beyond the "rules" I am demonstrating here. However, I think that physics enters deep enough into the basis for an artist's choice (whether he realizes it or not) that we should look at the implications, if only as a sort of intellectual game. The real point is that it is easy to predict mathematically, and show experimentally, that there are certain special combinations of sounds which have a particular effect on our brains, an effect quite different from those produced by random combinations even when they are very nearly the same as the special ones. All musicians should, and most do, face up to these peculiar combinations, since these combinations are lurking in everyone's ear.

Having got my little speech out of the way, I want now to show some examples of what happens when a musical sound—that is, one having a vibration recipe made up only of a fundamental and its harmonics— of one frequency is played along with a second one of another frequency. We can perfectly well assume that both sounds are played soft, so no combination frequency is produced to complicate things. Suppose that the two notes A and B have very nearly the same fundamental frequency, say 200 cps and 205 cps, respectively; we hear a beat between the two fundamentals which varies from loud to soft and back to loud five times in a second. In addition to this 5 cps beat, however, we are aware of a 10 cps beat that arises from the beating of the 400 cps second harmonic of A with the 410 cps second harmonic of B, and we also hear beats at the rate of 15 cps, 20 cps, and 25 cps, which come from the beating of the pairs of third, fourth, and fifth harmonics. The higher harmonics will not grind and fight so prominently since their beat rates are getting too high to recognize. Most people on hearing this "almost unison" will not consciously notice the separate sets of beats, but will simply agree that the sound is rough, grating, and generally unpleasant. As we tune the two fundamentals closer and closer together, the rate of beating will decrease progressively. For example, when A is 200 cps and B is at 201 cps, the beating of the fundamentals will be too slow for most people to discern (1 cps), while the second harmonics will beat at a fairly recognizable 2 cps rate, and so on. The rough sound produced by the fast beats of the higher harmonics will have decreased in intensity, and altogether the combination sounds much smoother and more music-like.

Listen to a piano tuner at work on the unison between two of the strings assigned to a single note. As each of the lower harmonics is brought well enough in tune that its beats are not noticeable, the sound becomes more musical, and the beating that you do hear sounds higher in pitch, rising step by step through the series of harmonics until the tuner is satisfied. I myself find the tuning of such unisons on a piano or other stringed instrument very gratifying; the clouds of discord seem to be driven gradually away by the tuning hammer, until I am left with the serene blue sky of musical "perfection" (almost!).

Let us now imagine that our ideal, softly played, uniform strings are tuned so that their fundamentals differ by a factor of very nearly two in frequency. For example we can take A to have a fundamental frequency of 100 cps, while B is at 205 cps. Under these conditions the fundamental of A will be heard free of any beats, while its second harmonic (200 cps) will beat five times a second with the fundamental of B. The third harmonic of A will not have anything to beat with, but the fourth will beat at 10 cps with the second harmonic of B, and so on up the series. Even though only half of the components can beat in this two-to-one case, we still get a roughness that goes away progressively as the two fundamentals are brought into the exact relationship. Similar arguments show that the exact three-to-one relation between the fundamentals A and B will also be a case of zero beating, with only a third of the harmonics contributing to the beating sounds. It goes without saying that all the other integral frequency ratios between fundamentals give rise to quiet, smooth, sounds when they are accurately tuned, although the violence with which our ears protest a mistuned interval decreases as the pitch interval increases, since there are fewer and fewer harmonics available which can beat.

Figure 21 shows some particularly interesting frequency ratios that come about from the beating of harmonics. The first line shows the two-to-one (musical octave) relation between the fundamentals A and B, assuming A to be at 100 cps. Any pair of frequencies coinciding on the diagram will beat if there is not an exact octave relation between A and B. The second line in the diagram shows the three-to-one relation (musical twelfth), which has already been described.

We know of many special intervals within the octave as well. For example, let us ask what happens when the fundamental frequency of B is 3/2 higher than that of A (that is, B at 150 cps and A at 100 cps). This is shown as the third line of Figure 21,

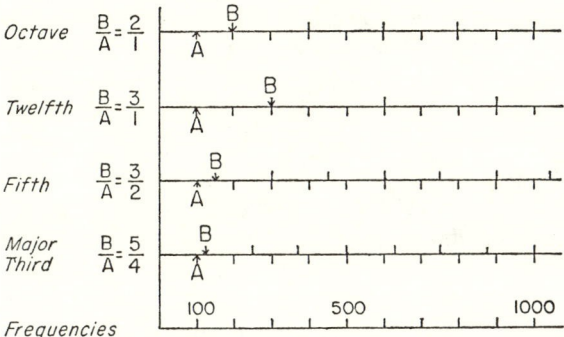

Fig. 21. The absence of beats between harmonics of two notes determines special musical intervals. In this diagram the arrows show the fundamental frequencies of note A and note B. The short vertical lines below and above each horizontal line show the positions of the harmonics of A and B, respectively.

where we discover that there are just as many pairs that can beat as we found in the second line of the diagram (once again it is the third, sixth, ninth, etc., harmonics of A which are eligible for beating). This new 3/2 interval is what the musicians call the "fifth," and constitutes one of the building blocks of harmony, taking its place right behind the unison and the octave in importance. The fourth line of Figure 21 shows another quite important pitch interval (the "major third"), in which the fundamental frequency of B is 5/4 that of A. There are not so many chances for beats among the harmonics here as in the earlier ratios, but enough remain to call attention to a mistuning.

Does any practicing musician really care about the relations we have found? The answer is emphatically yes. The foundation stone of harmony is the major chord, exemplified by the notes C, E, G, and C, arranged in ascending order. The frequency relations of these notes are just those we have been discussing: the C-to-E interval has a 5/4 frequency ratio, the C-to-G interval has a frequency ratio of 3/2, while of course C to C is the octave, or 2/1 ratio. Musicians have found that rearranging these three intervals, and moving one or another of them an octave up or down, will serve to define the whole musical scale and also to complete a system of harmony, the elements of which take on their musical distinctness from the presence or absence of beats among the harmonics of all the notes present.

I have not said anything about the influence of combination-tone generation of the ear on the special intervals used in music. It does not take a great deal of calculation to convince ourselves, however, that this "creative" function of the ear serves chiefly to strengthen even more the already strong hold of physics on musical form. You may find it interesting to work out some examples of this yourself, and to think about the way in which the rules of harmony can change with the loudness of the instruments playing it.

Before going on with our discussion of music, I should warn you that you will find all that I have said to be amply and clearly borne out by experiments you may do with violins, guitars, flutes, oboes, clarinets, etc. Indeed, practically any standard instrument *except the piano* will serve you well. Why this deservedly popular and versatile instrument is not suitable for *introducing* you to the theory of harmony is a paradoxical tale that will have to be told in the next chapter.

Harmony with Different Instruments

The sort of person who plays with an idea the way a monkey explores a new toy, turning it over, biting it, sniffing it, and seeing if it will survive a crash to the floor, is bound to ask whether it makes any difference to harmony if we use different musical instruments. Musicians will say yes, of course, but some of them may not know why it comes out that way. Anyone who has managed to come with me this far should be able to make up a plausible explanation, even though neither of us can compose symphonies or play them in public.

In order to save a little time in your thinking on this subject, let me propose a few experiments for you to consider. Suppose we ask whether a softly blown two-quart milk bottle (which sounds about an octave below middle C on the piano) has any special intervals connecting it with notes played on a violin. This last is an instrument whose vibration recipe is endowed with large amounts of *all* the harmonics of the note being played, but it cannot play a note low enough in pitch to get into unison with the bottle.

Now, a softly blown bottle produces only very weak harmonics of its own and none in our ears, and so we see with the help of Figure 22 that we should expect to find no special interval for this orchestration! Of course, a musician would say he recognizes the usual special intervals, but he does this with the help

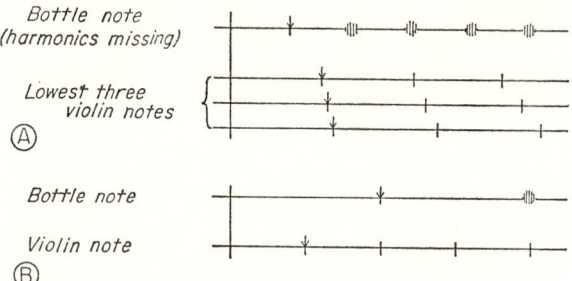

Fig. 22. *The relations between the note blown on a milk bottle and played on a violin are shown here. At (A) the bottle pitch is below that of the violin; no beating is possible, and there is no special interval. The hatched circles indicate the missing bottle harmonics. At (B) the bottle is filled with water to make it sound an octave above violin pitch. Slight mistuning will cause the bottle's note to sound with the violin's second harmonic.*

of his tonal memory, even when there is no physical effect to guide him. What the statement means for us at the moment is that we find no intervals that beat strongly when we play steady notes on the bottle while playing up the scale on the violin.

If we were really to try this experiment, we would actually hear a feeble beating sound in the immediate neighborhood of each special interval. It is easy to verify, however, that the beats are much weaker than in an experiment in which the bottle is filled with water until it is raised in pitch to a note somewhat higher than the B above piano middle C. This time the violin can get below the bottle frequency, and the strong violin harmonics can call attention to their presence by beating with the bottle fundamental. Figure 22 shows this sort of situation, when the violin is sounded one octave below the bottle's pitch.

Very little of the world's music is written for bottle and violin duet, so we really ought to look for an orchestral cousin to my rather outlandish example. Clarinets are unique among orchestral instruments in having a vibration recipe that contains only the odd-numbered harmonics strongly represented, along with the fundamental. That is, a clarinet playing the G below piano middle C has a fundamental frequency of about 196 cps, and its vibration recipe contains appreciable amounts of the $3 \times 196 = 588$ cps, $5 \times 196 = 980$ cps, etc., ingredients. For the moment we shall pretend that clarinets have *no* even harmonics at all, no ingredient 2×196 cps or 4×196 cps, etc.

Figure 23 shows what we get when two such simplified clarinets, *A* and *B*, softly play some of the special intervals we have already discovered. It goes without saying that the unison is special, since all the components present will beat strongly if one instrument is slightly mistuned relative to the other. Strangely enough, we find that the two-to-one frequency ratio (the octave) does not appear to be special at all, because there are no even harmonics to beat together in our simplified sound recipes! The interval of a twelfth, having a three-to-one ratio of fundamental frequencies, is special since some of the harmonics can beat if mistuned. Comparison with our earlier results, par-

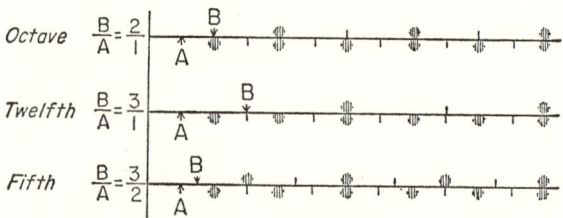

Fig. 23. *Simplified clarinets playing the notes A and B would have these special musical intervals. The circles indicate the locations of missing even harmonics. Notice that neither the octave nor the fifth can beat.*

ticularly the one shown in line two of Figure 21, tells us that the specialness of the three-to-one frequency ratio is stronger for clarinets than it is for ordinary instruments. As a matter of fact, it is as strong as is the octave interval between more normal instruments. When I speak of the strength of a special interval, it is just an informal way of saying how many of the harmonics can beat and quarrel to signal the presence of mistuning. The musical fifth, or three-to-two fundamental frequency ratio, is another interval which is beatless for our peculiar clarinets, as is shown in the third line of Figure 23.

Clarinets that have escaped from the simplifying hands of physicists, and that live in the real world of musicians, do have a certain amount of the even harmonics in their vibration recipes, so that beats can occur at all the spots marked with hatched circles in Figure 23. The result is that all the ordinary intervals remain special for a pair of clarinets, even though the "strength" of these is different from those found with more ordinary instruments.

We have been toying with a pair of spectacular examples which show that the tone color (timbre, vibration recipe) of the instruments played together alters musical affinities which we find between the notes they play. Such effects appear in greater or lesser degree throughout the orchestra.

A woodwind quartet made up of a flute, oboe, clarinet, and bassoon makes an almost gaudy little musical universe, if you want to play with possibilities in the way I have indicated. This is because the instruments have unusual properties:

> a) The flute has a vibration recipe almost wholly concentrated in the fundamental and first two or three harmonics.
>
> b and c) The oboe and bassoon have all the harmonics strongly represented.
>
> d) The clarinet, as we have seen, has its even harmonics weak.

You can have a lot of fun drawing diagrams for the sound resulting from an ordinary chord played with the different instruments taking various notes of the chord. If you are lucky, you can also try out these combinations with your wind-playing friends, or simply listen to recorded woodwind ensemble music, to get a feeling for the amazing array of tonal colors the composer has available to him. You are justified in suspecting me of being a player of woodwinds. I will not, however, allow you to discount my remarks about the colorfulness of woodwind music on the grounds of prejudice. Colorfulness and flexibility are not all one desires in music. For a steady diet I prefer to listen to a good string quartet.

Our initial explorations have shown us that music

does indeed rest heavily upon the physics of our ears; we have also run across hints of the ways in which instruments can be used to get various musical effects. We are now going to focus our attention on some of the main types of musical instruments to see how one goes about selecting vibrators which please our ears. In the next few chapters we shall visit these areas of Trans-Serbonia, and learn some of the native customs regarding the design and construction of various instruments.

CHAPTER VI

Stringed Instruments

Having learned a little how things vibrate and how our ears select certain combinations of sounds, we have been able to lay the groundwork for an understanding of the way in which music and musical instruments come about. In the last chapter we paid a brief visit to the regions in which harmony and melody are shaped, and we are now entering the lands inhabited by stringed instruments. I do not intend to take you on a tour of all the subtleties of such instruments. We shall look rather closely at two or three aspects of their behavior, and let the rest go.

We shall begin with the keyboard instruments, as represented by the clavichord, piano, and harpsichord. These all have their strings put in motion by sudden disturbances, after which they are allowed to vibrate freely until their energy is radiated away. As you know, the player of one of these instruments has no control over the pitch of his notes once the tuner has done his job. This means that all the special musical intervals must somehow be arranged beforehand by the tuner, a problem which cannot be solved except in a very ingenious but approximate way. After looking into the tuning of keyboard instruments, we shall move on to examine the violin family, in which the strings are kept in motion continually with the help of a bow. Once again I shall not describe the instruments in detail, but instead show you how the bow sets up the vibrations and controls their recipe, and how the properties of the body of the violin contribute to the sound we all expect from it.

The Strings of Keyboard Instruments

Almost everyone who has looked inside a piano has noticed that each note is supplied with at least one and sometimes two or three strings, whose lengths vary from about two inches at the treble end of the scale to a length of several feet at the bass end, depending on the size of the instrument. Let us see if we can find the physical reasons for this, and for the fact that on the piano, at least, the bass strings are heavily wrapped with copper wire.

The quantitative laws governing the vibrational frequencies of strings have been known for at least two hundred years, and may be summarized as follows:

a) A long string vibrates at a lower frequency than a short one under the same tension, in such a way that doubling the *length* lowers the pitch one octave.

b) The frequency of a string rises proportionally with the square root of the *tension*, so that one must quadruple the tension in order to raise the pitch an octave, or increase it by a factor of sixteen to raise it two octaves.

c) For fixed string tension and length, the frequency falls proportionally with the string *diameter*. (This is a way of bringing in the mass of unit length of string.)

Now let us "design" the strings for a piano. So we can learn why a piano is not in fact made this way, we shall start on the assumption that all the strings are to have the same diameter and tension as the customary ones for the top note. The highest C on a piano has strings almost exactly two inches long; so we find ourselves choosing string lengths of two, four, eight, and sixteen inches for the four highest C's on the keyboard. It comes out that the lowest one needs a string twenty-one feet long! Not only would our piano take up far too much room; we should also find that the tone color of the lowest notes would be rather bad unless, for reasons that we will learn later on, we used rather large hammers.

Our design is obviously a failure; we must try a different approach. For example, we might make all the higher notes according to the original design, shortening the lower ones so that they fit into the space we are willing to allow them, and then operating them at a lower tension than the rest. People who have tried this discovered that if the tension is cut in half, the tone quality becomes unexpectedly limp and bad.

Having thus been foiled twice, let us try our third and last possibility for maneuver: the diameter of the string can be increased at the lower end of the scale to make up for its deficient length. In desperation we measure all the string lengths of a grand piano and learn somewhere that the strings are all kept at a more or less constant tension, and upon this information we calculate the string diameters. All goes pretty well in the upper part of the scale, but once again something goes wrong at the bass end. The lowest "wires" we need would have to be little bars of steel nearly as thick as a pencil, and once again the tone we get is horrible. As if these troubles were not enough, we find that a wire of this thickness, length, and tension vibrates at a frequency considerably higher than our formula would lead us to expect.

During all these unsuccessful experiments, a physicist friend has been thinking of certain things in vibration theory which he does not always get around to telling his students, and which therefore he sometimes forgets. He reminds us that our ordinary formulas for vibrating strings are descended from our work on beaded strings. The string is supposed to be perfectly limp except for the rigidity it gets from being pulled tight; the restoring forces which act on the beads come only from the string tension. A real wire, on the other hand, has some stiffness in an amount that increases as the diameter is increased. Such a wire vibrates under the influence of two sets of forces, one set arising from the string tension, and the other from its stiffness. As a result the vibrational shapes and frequencies of all the various modes of oscillation are of a sort intermediate between those of a flexible string under tension and those of a stiff but unstretched bar whose ends are clamped. A little calculation shows that stiffness in a string raises all the vibrational frequencies somewhat, the higher modes being affected more than the lower ones. We can see now why our experiments failed; a slack string has its vibration recipe affected by stiffness. Its higher frequency components are not whole-number multiples of the lowest vibrational frequency, and it produces what we call an unmusical note. The pencil-sized rod we tried was also unmusical; here the tension forces were so small compared with the stiffness that the string thought of itself pretty much as a clamped bar, for which the first few vibrational frequencies are found at 1, 2.7, 5.4, and 8.9 times that of the lowest mode. A good musical string is then one in which the tension contributes as much (and stiffness as little) as possible of the forces which make it vibrate. Other things being equal, we want to use as thin and flexible strings as possible, and at the highest possible tension.

Now that we have tried our unsuccessful hands at piano stringing, and have learned the reasons for our failure, we can appreciate the way in which, over the years, practical men have solved the problem. Down to an octave below middle C, the strings are lengthened by a factor of 1.94 instead of 2 per octave, their diameter is increased by 9.3 per cent per octave, and the tension is reduced to the proper amount to bring the string into tune. Below this point the wires lengthen very little, and the pitch is lowered by using wires wound closely with copper wire in a way to add to the mass without increasing the stiffness unbearably. The diameter, and therefore the stiffness, of the last unwound string is chosen to match the stiffness of the first wound one, so that an even-sounding scale is obtained. We see from this little discussion that piano makers have had to back away from the ideal in order to get an instrument of a practical size, and that they have arranged things so that the heavily used middle of the piano is good, while the tone quality as you go down the scale is gradually spoiled in order not to advertise the not-so-good bass notes by setting them off as freaks.

Around middle C the notes of almost any honestly built piano can sound musical and clear, because the string proportions have not been compromised very much. At the bass end of the scale, however, anyone can hear the difference between the noble sounds issuing from a first-class concert grand and the clumping noises from the shrunken pianos sold as pieces of furniture. The reason is easily found: the bass strings of a concert grand are about twice as long as they are in a small piano, so that they can be pulled up to four times tension if the thicknesses are the same.

There is a partial escape from the troubles caused by a short, stiff string. Adding properly chosen weights to the string a couple of inches from one end helps to improve the tone quality somewhat, because such weights lower the frequencies of the various vibrational modes in a way which partially compensates for the raising caused by stiffness. These weights must be perfectly cylindrical if they are not to cause twisting vibrations in the wire. You can try this on your piano by wrapping a doubled foot-long piece of wire solder around the lowest string about two inches from one end of the vibrating length of the string. This will throw the note out of tune somewhat, but the change in tone quality will be interesting.

Before we leave the strings, we should notice the general size of the tension which is upon them, and also the magnitude of the force which the piano frame must withstand. On a concert grand the tension may go as high as 450 pounds per string, and the total pull distributed over the frame is in the neighborhood of twenty tons! With forces like these to contend with, a piano designer must be a good mechanical engineer as well as a capable vibration physicist. To be sure, pianos were built and built well before the days when people called in engineers and scientists to help them design machinery. Most of our present-day piano design was settled without help of formal technical knowledge. The evolutionary process was long and very slow, but its *results* can be summarized and passed along quite easily with the help of a little physics.

The Soundboard and the Bridge

If you were to stretch the best string in the world between two spikes driven into a concrete wall, and strike it at just the right place with a good felt hammer, the sound would be a faint and thin parody of a true piano tone. The string is such a narrow object (a few hundredths of an inch in diameter) and the wavelength of the sound it makes is so long (about four feet at middle C), that the wire is an exceedingly poor emitter of sound into the air. Because of this, piano strings are not stretched across a simple frame, but rather made to go over a wooden "bridge" that is fastened to the thin soundboard, in the manner diagramed in Figure 24. The usefully vibrating length of the string lies between the bridge (A) and the poetic-sounding capo d'astro bar (B), even though the tension is carried on beyond the bridge to a hitch pin (C) set in the steel frame at one end, and beyond the capo d'astro bar to the tuning pin (D) at the other.

When a string is struck, it tries to shake its fastenings with no success at the heavy metal of the capo d'astro bar, but the bridge is not so solidly made, and can be put into vibration. In this way the string is able to drive the large and elastically built soundboard (E), on which the bridge is mounted. Because of its size the soundboard is a good radiator of sound; it can take the string's vibrational energy and radiate it into the room. Because of the additional energy drain pro-

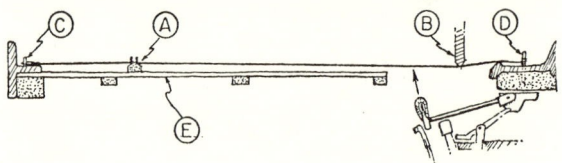

Fig. 24. The mechanism of a grand piano is shown here in cross section. The usefully vibrating length of the string is that between the bridge (A) and the capo d'astro bar (B). The hitch pin (C) and the tuning pin (D) bear the tension. The bridge, since it is attached to the soundboard (E), causes the soundboard to vibrate when the string is struck.

duced by the bridge, a piano string will die away in vibrational amplitude much more rapidly than a similar string that is mounted between two rigid supports. Thus it is incorrect to say that the soundboard amplifies the sound of a string, unless you mean that it enables the string to convert its energy more rapidly into sound and thus get rid of it quickly enough that only a little is dissipated in friction.

You will recall from Chapter III that any complex system has a large number of vibrational modes, and that an oscillating force of the proper frequency can drive any one of them to large amplitudes. In a piano the complete set of strings provides driving forces of a wide variety of frequencies, and we must be careful to adjust the resonances of our soundboard so that they do not clump into groups; otherwise certain notes will be radiated very differently from the others, and so give us an uneven-sounding scale. A nearly square soundboard without ribs or bridge would be very bad for a piano because of the grouping of its favorite vibrational frequencies, but a suitably placed set of ribs not only will strengthen it but also shift the resonances around. Fortunately for everyone, the odd wing shape of a typical soundboard, along with its irregularly spaced ribs, does not generally give much drastic trouble, but its design does help to make the distinction between a first-class piano and a mediocre one.

Vibration Recipes of Struck and Plucked Strings

Back in Chapter III I pointed out that the vibration recipe of a string depends on how it is started, and I gave some examples of the recipes produced by plucked strings. If you look at these recipes again, you will notice that plucking the string at a point one third of the way from one end gives a recipe in which the third, sixth, ninth, etc., modes are missing, while plucking it at a sixth of the way along gives a recipe with the sixth, twelfth, eighteenth, etc., modes missing. In 1800 Thomas Young[1] discovered the regular-

[1] Young is most famous for the work he did to establish that light is a wave motion, and did not consist of a stream of particles as Newton thought. In recent years we have come to realize that light is a kind of radiation which possesses both wave and particle properties, but this neither vindicates Newton nor repudiates Young. We would sadly miss the contributions of both to physics.

ity that I am hinting at, and stated it in a more general way that applies even to non-uniform strings. No mode of vibration can be set up that has a node (stationary point) at the position at which the disturbance is applied.

These properties of a plucked string allow us to understand how it is that the placement of the harpsichord "jack" can have such a big effect on the tone quality. Practical instrument builders learned this with amazing slowness. Harpsichords were built with incorrect jack placement long after people had come to recognize the superiority of certain instruments over others. Sad to relate, there are a few contemporary harpsichord builders who have let their eagerness for "authenticity" override their common sense to the extent that they copy an inferior instrument instead of letting our modern knowledge help them achieve what a few ancient builders got as a rare and much-prized accident.

Different plucking positions "suppress" different sets of frequencies, and as a result provide our ears with different kinds of sound. It is often said that a harpsichord is "expressionless," meaning that the timbre and loudness of a played note cannot be altered by the player. Composers for this instrument had to accent important notes by elaborate trills and ornaments and rapid arpeggios. In a partial sense this is true, but not completely. In a rapidly played harpsichord the key is not pressed gradually until the string slips off the quill; more usually it is snapped down rapidly, so that the starting string shape is a little different from the simple triangular shape, as shown in Figure 25. The slightly sagging shape gives a vibration recipe that differs from that of a slowly plucked string chiefly in having larger amounts of the higher frequency components; it produces a tone color that is a little "brighter" and "more penetrating," as the musicians say. If you can get your hands on a harpsichord, it is not hard to convince yourself of the different sound quality produced by a slow pressure on a key as compared with that coming from a sharp,

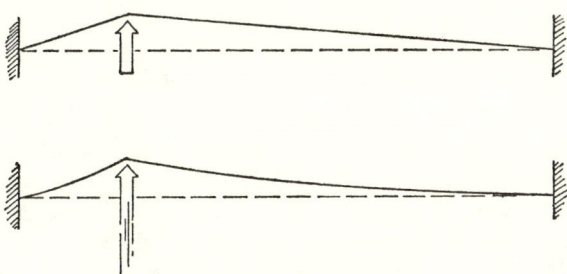

Fig. 25. Whether a key is slowly depressed or hit a sharp blow affects the harpsichord's tone quality. A slowly depressed key gives the string the triangular initial shape shown in the upper diagram. A sharp blow on the key starts the string off with the slightly sagging shape of the lower diagram, and alters the tone quality.

quick blow. You can hear the same effect, although not so clearly, in recorded harpsichord music.

I now mention the clavichord, not so much because of its musical importance as for the interesting physics of its intermediate position between harpsichord and piano. When you press a key on the clavichord, a little steel wedge called a "tangent" comes up and displaces the string a short distance, keeping it displaced for the whole duration of the note. We have here a new way to start a vibration, in which one of the fixed ends of the string is suddenly shifted to a new position. The resulting initial shape is somewhat like that of a string that has been plucked very near to one end, and it turns out that the more quickly the key is depressed, the nearer to one end the string appears to have been plucked. In addition to the change in loudness produced by varying degrees of vigor in playing, there is also an alteration in the tone quality. In such an instrument the tangent divides the string into two parts which could both vibrate if the part nearer the tuning pins were not damped permanently by a strip of felt. The sound of a clavichord is somewhat like that of a harpsichord, although much less loud. One of my friends said that it reminded him of the noise made by hairpins falling on the floor. Nevertheless, there is a certain charm to clavichord music, and a real advantage to it if your landlord hates to hear you practice.

As you know, a piano string is set into forceful vibration by means of a hammer blow. Beyond reminding you of Young's law, I need say very little

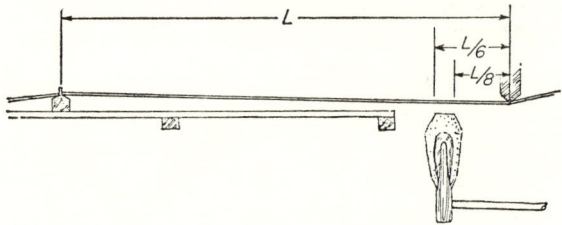

Fig. 26. The very broad hammer in this imaginary piano string arrangement would kill off many vibrational modes. In a real piano most of the hammers are considerably narrower than this in proportion to string length.

about the effect of hammer position on the kind of tone we get. During the hammer blow a certain length of string is in contact with the hammer face, so a *range* of harmonics is omitted from the vibration recipe. Suppose that the hammer is arranged to touch the string in the interval extending from about an eighth to a sixth of the way along the string, as shown in Figure 26. The following table was constructed for such a string with the help of Young's law, and gives a list of the string modes that do not contribute to the vibration recipe.

MISSING	MODES		
6	12	18	24
7	14	21	28
8	16	24	32

If you arrange these in order, you discover that the vibration recipe has been deprived of a large part of its possible set of ingredients. Our string will give a very different sound from that which it would make under the stimulus of a narrow hammer. You can now see why it is that the hammers of our original attempt at piano design would have to be big. If we took our proportioning literally, and chose the width of contact of the highest hammer to be about an eighth of an inch, we should find that the bottom note would need a hammer that touched the string over a length of sixteen inches. It would take a muscular virtuoso indeed to play rapid music on such an instrument!

We find one more source of complication in the vibration recipe of a piano string. When a hammer strikes the string, it stays in contact with it for a short time, during which the parts of the string on either side of it are free to vibrate in a way that is exactly like the vibrations of a clavichord string. The frequencies of these new vibrations are related to the lengths of the two parts of the divided string, and not to the length of the complete string, and they do not necessarily have a musically simple relation to the desired note. These clavichord vibrations not only contribute to the forces which cause the hammer to rebound, but they also can contribute quite an unmusical clang to the tone if they are not damped out sufficiently by the felt hammer head. For a piano tuner it is a matter of considerable practice to adjust the hardness and breadth of the striking area of all the hammers in a piano in a way that gives a uniformly good succession of sounds as the scale is played.

Up until now our wanderings have shown us a number of mechanical properties of the piano which must be correctly arranged in order to get a musically satisfactory sound, and lest you get the impression that the player has nothing to do but strike the proper notes, I would like to say a little about "touch." Piano actions are constructed in such a way that there is no connection between a hammer and its key during the last part of the hammer swing. No matter how subtly you graduate your "touch" upon a given key, all that you are doing is setting up a certain velocity in the hammer, which then flies freely over to where it strikes the string. As you well know, a gentle push exerted for a long time is equivalent in every way to a short, hard shove, as long as they both give the hammer the same final velocity. The string has no way of knowing whether its hammer originally got its velocity from the most skillful of pianists or from being shot from the muzzle of a popgun—it will give out exactly the same sound. A musician can learn to choose how hard he strikes a given key for the best musical effect, and if he wants to think of himself as "shaping" the note by an elaborate variation of pressure on the key as he depresses it, let him. What he is doing is a sort of dance which helps him to relate the various parts of his music into a coherent whole.

Now that I have baldly stated that the tone color and loudness of a single note are unaffected by the *way* in which a key is pressed, and depend only on the hammer velocity at the time of striking, I must hasten to add that the musician does indeed have the ability to alter the kind of sound he gets when playing a chord. The vibration recipe we get when a number of notes are played together is made up of the ingredients supplied by all the separate strings that are in motion, so that the resulting recipe depends on the amounts contributed by each string. As a result, the sound we hear depends for its timbre, not only on the particular strings which have been struck, but also on the loudness of each note of the chord relative to the others. A pianist also can alter the tonal effect in melodic passages by varying the amount of overlap between successive notes, thus producing more or less noticeable consonances and dissonances. The "loud" or sustaining pedal is also used for this purpose on a much larger scale. In addition, raising all the dampers with the loud pedal permits certain of the strings to be driven into "sympathetic" vibration; that is, any vibrational mode of the unstruck strings that happens to match some part of the vibration recipe of the struck strings will be excited, causing a change in tone color. Through long practice the musician has learned to choose the kind of sound he wants, and how to get it from the piano, even though he may be completely unconscious of the actual manipulations he uses to get his effects. It is incredible how skillfully a man can learn to exploit the possibilities of a complicated musical instrument—and usually without having much of an idea of how he goes about it.

Tuning the Piano: a Dilemma

In the last chapter we learned some of the guiding principles of music that are laid down by the construction of our ears. We found that certain relations among the frequencies of vibrations we hear are distinguished from all others by having a smooth beatless sound which turns quite rough if the relation is altered just a little. Most musical instruments are constructed in a way that permits the player to make small adjustments to the pitch of the notes he plays, so as to satisfy the changing requirements of the music. In a reasonably well-designed instrument he can do this in spite of any inaccuracy which may have come into its construction. On keyboard instruments no such adjustment is possible for the player; he must be careful to get all the notes properly tuned beforehand. It turns out that this is a practical impossibility, but that a satisfactory compromise can be found.

European music is hung upon sequences of notes arranged in ascending order, which are called scales. A scale, for reasons we have met earlier, is defined not so much by the exact frequencies of its notes as by the frequency ratios between successive notes. Suppose we start out with a familiar scale, one that is got by playing middle C on the piano and then each of the succeeding white keys above it in pitch, for one octave. This sequence of pitches ultimately derives from the numerical relations favored by our ears, and we can set down in a table the ratio between the frequency of the starting note C and that of each succeeding note in the scale. As I warned you earlier, the piano is not tuned exactly to the intervals given here, but to a close approximation.

C	D	E	F	G	A	B	C
1.000	1.125	1.250	1.333	1.500	1.667	1.875	2.000
1/1	9/8	5/4	4/3	3/2	5/3	15/8	2/1

This scale can be extended upward into the next octave by simply doubling all the numbers, and downward an octave by halving them all. A piano tuner could adjust all the white keys on a piano to this sequence of pitches, and you could play many different kinds of simple music on the notes he has made available to you. Suppose, however, you decide to play a simple melody which originally began on C in a new way by starting it on the next higher note (D) of the scale. The result will be somewhat odd; it certainly will not sound like the original tune, and your listeners will swear that you have played some wrong notes. *Yankee Doodle* is a good familiar tune to try this on. Play only on the white keys of the piano, beginning first on C and then on D. Since the tuner is still around, and is willing to tinker, you ask him to adjust some of the leftover strings in the piano so you can play your melody beginning at D, but with some of the newly tuned strings used as replacements for those in the original C scale which were at the wrong pitch. Once this has been done, you can start any of your simple music at either one of two places on the keyboard, and play it successfully. In addition to this small kind of freedom, you also find that there are interesting harmonies to be made by combining notes from the new D scale with those from the original C scale, harmonies which can be traced back as usual to the special pitch intervals we met earlier. The additional musical possibilities implied by this discovery are so entrancing that all thought of common sense and practicality disappears, and the poor tuner is asked to supply you with a set of strings which gives you a properly acceptable scale beginning on every one of the original notes of the C scale. A little thought and a lot of arithmetic on the part of the tuner unearth the fact that each octave would require nineteen keys, if we include the C's at both ends. A whole new piano would have to be built with many extra sets of strings. Such a new tuning would give you so many interrelations between notes that great new vistas are opened to your musical creativity, so you place an order for the job. When you try out this latest musical creation, you discover that in several places on its complicated keyboard the pitches of adjacent notes are so close together that it hardly matters which key you press—the harmonies come out almost as well either way. You also discover that several notes in the new scale are used over and over in most of the music that you play. On the other hand, it is still necessary to play wrong notes in certain circumstances, in spite of the fact that the tuner has arranged to give you more than twice as many notes in an octave as you started with.

I have been making you act in a rather farfetched little drama, but it has a meaning for us. If we wanted to play arbitrarily complicated music and do it in perfectly accurate tune so that all our special intervals were strictly adhered to, we would need a piano with hundreds of notes in each octave. On the other hand, we find that for the vast majority of music that anyone can compose, the needed notes group themselves into clumps. That is, we find groups of adjacent strings that are tuned to very nearly the same pitch, so that we are tempted to replace them with a single string, tuned to a judiciously selected intermediate pitch that will produce tolerably slow beats whenever it is used in a chord. Over the last two hundred years dozens of these compromise tunings ("temperaments") have been devised, with varying numbers of notes in an octave. Some of these were quite workable, and some much less so, and it took the musical authority and skill of men like J. S. Bach to convince the musical world that one of the simplest of these systems can provide a good practical basis for keyboard music.[2]

This compromise system, which goes by the name of "equal temperament," divides the octave up into twelve parts in such a way that each note of the rising sequence is a fixed ratio higher in frequency than its immediate neighbor. Suppose we call this ratio $K,$ and start our scale at a note which has the frequency $f_0.$ The next note will have a frequency f_1 that is equal to $Kf_0,$ while the next one above this will be at a frequency $f_2 = Kf_1$ which is the same as writing $f_2 = K^2f_0.$ If we follow this pattern, we find that the

[2] The details of how different temperaments came about are simply and clearly explained in Alexander Wood's *The Physical Basis of Music,* and in L. S. Lloyd's *Music and Sound.* Helmholtz' *The Sensations of Tone* is well worth browsing in as well, since it gives an exhaustive treatment of the subject.

sixth note above the lowest has a frequency that is K^6 times f_0, the seventh one K^7 times f_0, and so on. The twelfth note is supposed to be an octave above the first, and therefore should have a frequency that is two times f_0. Because of this, we find that the twelfth power of K has to be two ($K^{12} = 2$), and therefore K itself is the twelfth root of two, or 1.0595. The sequence constructed in this way, with each note about 6 per cent higher in frequency than its predecessor, is called the even-tempered chromatic scale, and all the other scales we want are constructed by choosing a definite sequence of single and double skips up among these intervals. For convenience the pitch interval between successive notes in the chromatic scale is called a semitone, and wider skips can be described either in terms of the number of semitones they include, or by names that are more directly related to their musical uses.

In the even-tempered tuning, the simple C scale with which we started is approximated by the following set of intervals between successive notes:

C	D	E	F	G	A	B	C
tone	tone	semitone	tone	tone	tone	semitone

On a piano keyboard the tuner does not make any difference in tuning between the black and white keys —they are all arranged in a uniformly rising sequence of pitch. The two colors and shapes of keys are there only to help the player to find his way by feel over the wide and otherwise trackless expanse of the keyboard. Except for the small musical effects caused by differences in leverage between the longer white keys and the shorter black ones, a pianist can start any piece of music he pleases on any note he pleases and have it sound equally good, the approximations brought in by the system of temperament being identically the same for all beginning places. Because of this, it is possible to make a table of frequency ratios that can be used to construct a chromatic scale from any desired pitch, although nowadays it is customary to begin at the A above middle C, tuning it accurately to a frequency of 440 cps. The following table gives the even-tempered frequency ratios for a chromatic scale, along with the ideal ratios that we got directly from first musical principles, and the actual frequencies to which the octave above middle C is normally tuned.

NOTE IN SCALE (BEGINNING AT MIDDLE C)	EQUAL TEMPERAMENT		IDEAL TEMPERAMENT
	RATIO	FREQUENCY	RATIO
C	1.0000	261.63	1.0000
C♯	1.0595	277.18	
D	1.1225	293.66	1.1250
D♯	1.1892	311.13	
E	1.2600	329.63	1.2500
F	1.3348	349.23	1.3333
F♯	1.4142	369.99	
G	1.4983	391.99	1.5000
G♯	1.5874	415.31	
A	1.6818	440.00	1.6666
A♯	1.7818	466.16	
B	1.8877	493.88	1.8750
C	2.0000	523.25	2.0000

It is time now to look at our approximation to see that we have not sold ourselves too cheaply. A glance at the table and a little arithmetic show us that the musically important 3/2 ratio (C to G, C♯ to G♯, D to A, etc.) is in error by only .41 beats per second in the middle octave of the piano, while the worst of all the intervals of the even-tempered approximation (the 5/4, or C-to-E interval) beats at the rate of 2.6 a second. When you consider that the loudness of a piano note has fallen away a great deal in two seconds, it is apparent that the slow beat of the 3/2 interval will hardly be detected, while the 5/4 interval gives a wavering sound which, played in the middle of the piano range, is not terribly obtrusive to most ears. The beating rates for our inexact intervals increase as we go higher up the scale, and, as a matter of fact, double for every octave that the pitch is shifted up. These faster beats might be more noticeable than they turn out to be if it were not for some other properties of the piano.

We have had to worry already about the effect of string stiffness on the tone quality of sound from a single string, but we are now concerned with the behavior of two strings sounded together. Suppose we try to tune two strings so that they are an octave apart by adjusting them to "zero beat" while playing them softly. This procedure is one in which the second vibrational mode of the lower string is put in unison with the first mode of the upper string, and does not give us a true octave relation. Stiffness in the strings makes the higher modes all a little sharper than the whole number ratios would indicate, and as a result tuning a succession of octaves up the piano keyboard has a tendency to run everything sharp at the top. If, on the other hand, the keys are struck hard during the tuning, it is possible to find not one but two closely spaced frequencies where the beating is least. The higher pitched of these is due to the process we have just discussed, while the lower one comes from a beat between the fundamental of the upper string and the true second harmonic generated by our ears from the fundamental of the lower string. Although it is sometimes hard to pick out these two intervals and hear them separately, it is quite possible to tune a piano using either one, as I can vouch from personal experience. The two points of least beat can be distinguished only at the high end of the piano (they run together in the middle range), but in tuning we need worry only about detectable effects. No matter which way the tuning is carried out, there will be an additional roughness brought into the tone by the string stiffness, and I have really no idea which is preferable from the point of view of general piano tone. On the other hand, a piano tuned with "unstretched" octaves (using the ear's own generated harmonic) definitely makes life easier for a violinist or clarinetist who wants to play sonatas with the piano.

I have enumerated several reasons why the music played on a piano must of necessity depart from the ideally calculated relations, but even so, all these "errors" add up to only a small alteration in the musical effect. Chords on a piano tuned to equal temperament sound a little more "bright" and "metallic" than they would on an instrument giving theoretically exact pitches, with no musical harm done. Composers have learned to make a virtue of the apparent fault, and can write music that is very successful when played on a piano but often sounds dull and insipid

131

when transcribed for other instruments.[3] The fundamental reason for the acceptability of even temperament is that the slight roughening of chords is in practice a small price to pay for the increased musical versatility it buys for keyboard instruments. Most of the other instruments are built in such a way as to keep their own inherent tuning as close as possible to the even-tempered system. Yet when one of these is used along with a piano, a skillful player sometimes will unconsciously depart from even temperament for the purpose of getting a more musical compromise. This necessity for continually adapting to the piano's even temperament puts somewhat of a burden on the instrumentalist who plays with the piano. The problem is immeasurably increased if the piano has not been tuned recently, for then the hapless player has nothing to relate himself to harmonically. Musicians away from the piano in a good ensemble will use the inherent pitch flexibility of their instruments to play with practically ideal temperament.

Bowed Instruments: the Violin

We are all familiar with the general appearance of the violin, which is a peculiarly shaped box on which four strings are stretched so that they run over a bridge that couples their vibrations to the box and its enclosed air. We also know that sounds are brought out of this device by rubbing the strings with a bow,

---

[3] A good piano freshly tuned by an expert has a beautiful soft smooth sound in spite of its unavoidable compromises. Alas, it takes only a few days' playing for the bloom to wear off and for the tone to begin its slow but steady change toward the harshness of a barroom piano.

and that the player chooses different pitches by pressing down with his fingers on one or another of the bowed strings in such a way as to shorten its vibrating length. Having summarized most of the familiar things about violins, I want to take you on a short tour of some of the physics involved.

The Bow and the String

We all have sat in a classroom when the teacher caused the chalk to screech against the blackboard, and some of us have noticed that a screeching chalk always draws a finely dotted line. Let us look into the way in which this dotted line is formed, since it is very similar in its mechanism to the one by which a violin bow works, as well as to the chattering of automobile clutches, the squealing of brakes, or the squeaking of really clean, long wet hair under a comb. From the point of view of physics, the chalk is a small block of matter mounted on a spring provided by our finger tips so that it forms part of a vibrating system. As it is rubbed along, the tip wedges against the board and comes to rest. The other end is still moving steadily however, so that elastic energy is stored in the spring, and the force trying to break the tip loose increases steadily until the frictional force is no longer enough to hold it. At this instant the tip flies forward in the beginning of a simple harmonic oscillation, using the energy that was stored in the spring. On the return swing of this simple oscillation the chalk again digs into the board and is brought to rest so as to prepare it for its next jump.

In all such "stick-slip" oscillators the frequency

with which vibration actually occurs depends chiefly on the spring stiffness and the mass of the moving object, but it can also be affected by the friction between the vibrator and its exciter, especially if the amplitude of the oscillation is large. The whole phenomenon ultimately requires a friction force which is greater between objects at rest or slow-moving with respect to one another than it is for objects that are sliding on each other relatively fast. Rosin is rubbed on the horsehair of a violin bow to give us this desired sort of friction (it is sometimes called "dry friction"), while a little oil or grease would kill off the vibration at once, since an oily surface is one in which the friction is least instead of most when the surfaces rub slowly. Even heavy grease on the string will not make a violin sound when bowed, while a smooth dry stick can excite string sounds when used on a violin. Thus we see that it is not the *amount* of friction, but its *kind,* that decides whether oscillation can occur.

The vibration recipe set up by a violin bow on a string cannot be calculated as easily as that of a struck or plucked string where there is only one initial disturbance, but there are a few general principles which we can set down. As in the case of the piano, we can invoke Young's law, which tells us about the higher modes which have nodes in that part of the string in contact with the bow—none of the modes is excited. If a violinist were to play with a bow carrying only a single hair, we should find that the vibration recipe lacked a single set of modal frequencies, as would a piano string struck with a narrow hammer. On a real violin, however, the player has his choice of where he will bow and also of the amount of bow hair that he will put in contact with the string; in the piano this is equivalent to a movable hammer with adjustable width. There is another way in which the bow can affect the tone quality. A violin string is quite heavily damped, and it can be shown that any vibrational modes which have nodes close to the place of bowing are only weakly excited. As a result the presence of damping superposes its qualitative effects on those of a wide bow, and the number of vibrational modes that are discriminated against in the bowing is increased still further by an amount that can itself be varied by the bow speed and pressure! You can verify my statement about the heavy damping by noticing how much more quickly the sound of a plucked violin string dies away than that of the same note played on a guitar or piano. A bowed string is additionally damped by the residual friction between string and bow during the "forward" jump of the string in its cycle of vibration.

All these phenomena help us to understand how it happens that a violinist can exercise such enormous control over the kinds of sound he can get from an instrument once he has mastered the musicianly handling of his bow. The history of violins has many stories about a skilled musician's playing a pair of instruments alternately, while a committee in the next room sits in judgment on their comparative quality. It is astonishing how varied some of the constructions are that can satisfy even a trained listener in a test of this sort, but it is always done with the help of a conscientious musician who is playing his heart out to get the best from both instruments! Thus we see that the real test of a good violin is not in the ears of the listener, but rather in the hands of the player. If he can get the range of effects he wants, without strain, then he has a good instrument. Except in a minor way, by clearing away some of the most obvious stupidities, physics has not been able to help him very much more than has the witchcraft that has grown up lush and rank around this beautiful instrument.

The Body of a Violin

We learned in Chapter III that any complex vibrating system can be analyzed into a set of equivalent simple oscillators, any one of which could be started up independently of the others. We have made considerable use of this possibility, and we shall use it a great deal more, but for a while it will be convenient to divide our violin into three main subsections, whose building-block oscillations are then imagined to "drive" each other in the manner that a child drives his swing. Already we have been looking at one of these systems, the strings, as though the rest of the violin did not exist. Another sub-system will be the wooden box itself, and finally we have the air that is enclosed within the box. The string is, of course, the only one of these sub-groups that has enough simple properties for us to examine its special qualities in a systematic way. We shall confine ourselves to brief remarks about the others.

While the air in the box is odd-shaped, it is not a "lumpy" system like a string of beads, and we can be sure that it has an extremely large number of resonant frequencies. If the motion of the wooden enclosure is at one or another of these "cavity resonance" frequencies, then we can be sure that the air will find itself in violent oscillation, and as a result it will radiate strongly into the room by way of the "f-holes" cut in the belly of the violin. You can find some of these resonances quite easily yourself if you sing softly into one of the f-holes as you go "ooooo" (as in shoot) and slide upward steadily in pitch from the A below piano middle C on up to a pitch two octaves higher. You will hear and also feel a good strong resonance just above middle C, and perhaps find others higher up near F♯ and A, and finally a loud but very narrow resonance about an octave above middle C. Before doing this experiment it is a good idea to wedge a handkerchief under the tailpiece to damp out its resonances, and also to wrap your fingers lightly around the neck of the instrument to kill off any distractions from the strings.

A crude experiment of this sort is not enough to tell which of these resonances is due to the cavity and which due to the wood itself, but it is possible to settle the matter easily for the lowest of them. Partially closing an f-hole will markedly lower the pitch of this lowest resonance, and the lowering depends on the amount of hole covered and not on the weight of the cover. This shows that it is the air motion that is being altered and not the mechanical vibration of the wood.

The string providing the original driving vibrations is coupled to the box and its air by way of the bridge, a device which itself forms a spring-mass system with a set of resonances. The lower part of Figure 27 shows the bridge (*A*) as it is mounted on the belly of a violin. Before you get to the wide curving top part

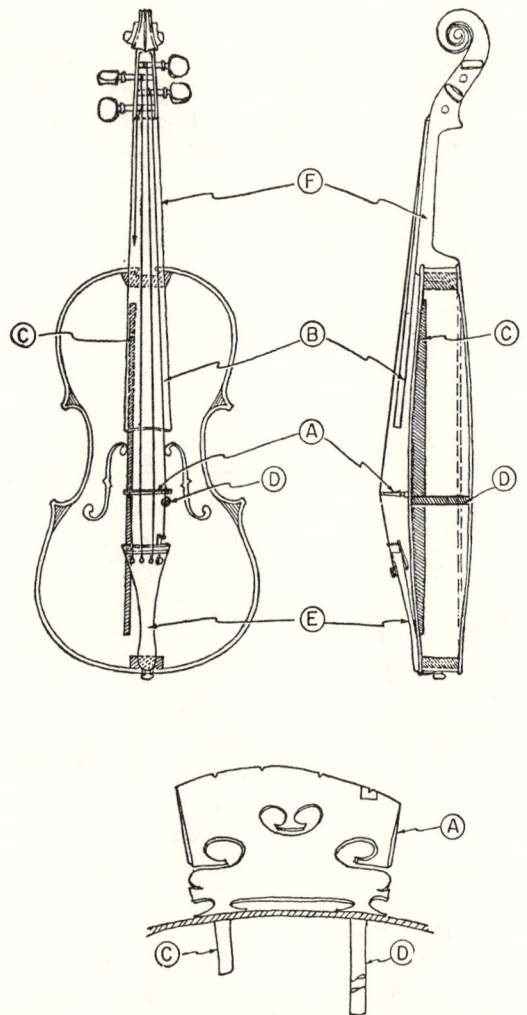

Fig. 27. The inner workings of a violin can be studied with the help of these drawings. The feet of the bridge (A) rest on the bass bar (C) and the sound post (D). The fingerboard is at (B), the neck at (F) and the tailpiece, or string holder, at (E).

on which the string rests, notice that the bridge has two widely spaced feet and then a thinned-in waist. Careful measurements have shown that the top of the bridge vibrates back and forth on its slim waist when it is driven at certain frequencies, and thus alters the forces which it passes on to the violin belly in a way depending strongly on frequency. Let us turn now to the upper two views of a violin in Figure 27. These are drawn as though the instrument were made of glass, so that the inner workings can be seen as well as the outer parts. The bridge (A) stands so that one foot is over the long narrow "bass-bar" (C), which extends most of the way along the belly on the left side, while the other foot is near a thin pencil of wood called the sound post (D), which is wedged between the thin and flexible belly and the more rigid back piece. Bowing a note produces a side-to-side force on the bridge exerted by the string, and this causes the bridge to rock back and forth about the sound post so that the bass-bar can set the whole belly into a sort of twisting oscillation.

A great deal of the skill of a violin maker is to arrange all the various kinds of vibration that are possible in the instrument in a way that gives the characteristic sound people want, and at the same time to make sure that none of these resonances is too strong.

Violinists are plagued by what they call a "wolf note," which is connected with the resonance that I

Note in the lower drawing that the bridge's shape is unsymmetrical, to locate the strings conveniently for the player, and that it is cut away to give it the desired resonant frequencies.

suggested you look for an octave above middle C. The damping of this resonance is very small. It responds strongly to a properly tuned driving force, and on a poor violin the natural frequency of this vibration may coincide with one of the notes of the musical scale. Whenever the musician trys to play this note, he finds that his instrument has a voracious appetite for vibrational energy, and this note behaves very differently under his fingers from any of the neighboring notes of the scale. A closer look at the physics of the situation tells us that where two parts of a complex system have the same natural frequency when taken separately, they affect each other so strongly when coupled that all sorts of new phenomena occur. The most prominent of these effects is a sort of beating in the vibration; the played sound comes out harsh and wavering. (I would like to emphasize that the phenomenon I am describing here has only a superficial connection with the beats that we made such great use of in the last two chapters.) The fact that the bow is working on only part of the coupled system brings in a whole set of other complications, and on top of these comes one that is due to the stick-slip action of the bow friction: you sometimes can get wolf-note effects by playing a note whose pitch is an octave or a twelfth higher than the box resonance. We shall meet a very important and useful cousin of this phenomenon when we learn about the "privileged notes" of a trumpet.

Fortunately for violinists and violin makers, the very strength of the troublesome wolf-note resonance in the box also makes it an extremely narrow one, as we learned in Chapter III. As a practical matter the effective width of the resonance is slightly less than a semitone, and a little judicious tinkering with the position of the sound post or bridge or bass-bar can usually smuggle the offender into the space between two notes of the chromatic scale.

While I have not said anything specific about the other members of the bowed string family, you will certainly recognize that violas and cellos show many similarities to the violin in their construction and nature. Because of the flexibility of tone and pitch and loudness all these instruments possess, they have always been favorites of musicians. The two violins, viola, and cello of a string quartet together make a combination whose possibilities have attracted and challenged the best composers for generations. Even so, its apparent simplicity leaves no hiding place for an unskillful phrase, nor can anyone cover up a weak musical idea with bombast in the way you occasionally find in symphonic composition.

CHAPTER VII

Vibrations in Pipes and Horns

Our travels so far have led us in and around the reasons why music takes on its regularities, and have also shown us how and why uniform vibrating strings are good basic devices for musical instruments. We are now preparing to turn our attention to the wind instruments, which are much more varied and complicated than are the strings at the level of our present inquiry. Before we do this, however, we have to learn a few more items of physics, a process which we have gone through several times. We must prepare ourselves for travel in new territories, and it is my duty as your guide to suggest the weapons we require, and also the beads we need for bribing the natives into telling us their secrets!

Bottles, Balls, and Springs

The air in a wind instrument must be in vibration, or it could not radiate sound out into the room. We know that air is made up of molecules; the musical air column is a complex vibrator having many, many modes of vibration. We know that wind instruments are silent unless someone blows upon them, which shows that these vibrational modes are driven by energy put in from the outside in some fashion. Although these bare-bones statements are quite obvious to us, we must put some flesh on them before they can tell us how to proceed.

The behavior of any vibrator depends on the forces acting between its particles; for example, each particle of a string feels forces (exerted by its nearest neighbors), which are more or less at right angles to the general line of the string. That is, the vibrations we have discussed so far have all been from one side to the other of the line of the stationary string, in what we call *transverse* vibration. Air molecules, on the other hand, act very much like smooth elastic balls which can exert head-on pushes on one another, but no sideways "shearing" forces, and air motions in a pipe or horn occur in a direction parallel to the length of the instrument. We refer to such motions as *longitudinal* vibrations.

Our introductory vibrator for air vibrations will be the gallon jug, which will replace the simple pendulum with which this book began. If we pull out the cork sharply, we hear a hollow ringing sound, which indicates the presence of a vibration in the bottle. It does not take a great deal of imagination to see that air is rushing in and out of the neck at the vibrational frequency. Let us look at this a little more closely, supposing meanwhile that the air in the neck has solidified into a smoothly fitting piston which can slide up and down the neck without allowing any of the air *in* the bottle to leak out. The air trapped in the jug serves as a sort of spring upon which the neck piston can bounce with a motion which is simple harmonic for all tolerably gentle excitations. We can see this with the help of Figure 28, which shows, for example, that the pressure inside the bottle increases as the piston moves downward; the net force on it (from air pres-

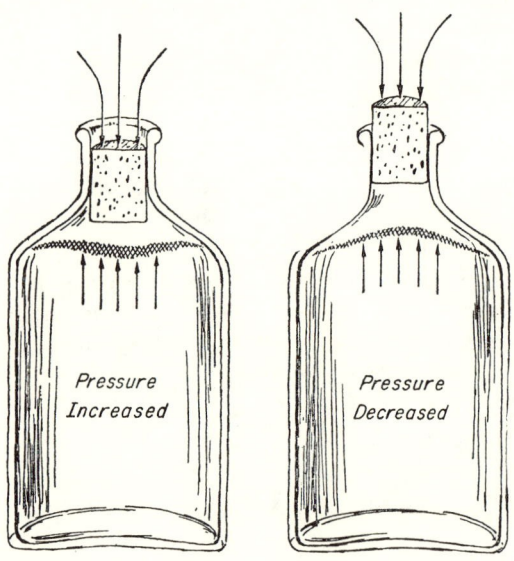

Fig. 28. Stoppered bottles can act as oscillators. If the slug of air in the neck is below its equilibrium position (at left), the pressure inside the bottle increases, and there is a net upward force on the "piston." Conversely, a piston above its rest position (at right) reduces the pressure in the bottle, and the atmospheric pressure tends to push the slug downward. Either way, there is a restoring force opposite to the displacement. Simple harmonic motion is possible.

144

sures exerted on its top and bottom) is upward, providing a restoring force as is needed for simple harmonic motion. Helmholtz was the first man to clarify our scientific thinking about these "cavity resonators," and as a result they are often called by his name. Lord Rayleigh did some very elegant experimental and mathematical work on the subject, which he describes in a rather technical book entitled *Theory of Sound*.

Figure 29 serves to call our attention to the fact that the size of the bottle has a great deal to do with the frequency of oscillation; a large volume acts like a weak spring, and a small one has the properties of a stiff spring.

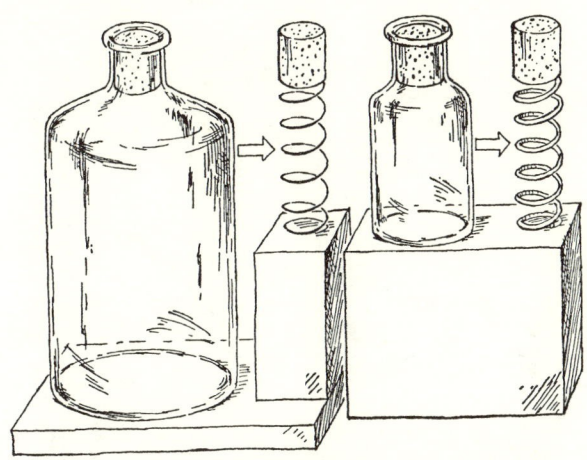

Fig. 29. If you think of the air in a bottle as a spring acting on a piston (the movable stopper or plug), you will see why the size of the bottle has a great deal to do with frequency of oscillation. A large volume acts as a weak spring, a small volume as a stiff spring.

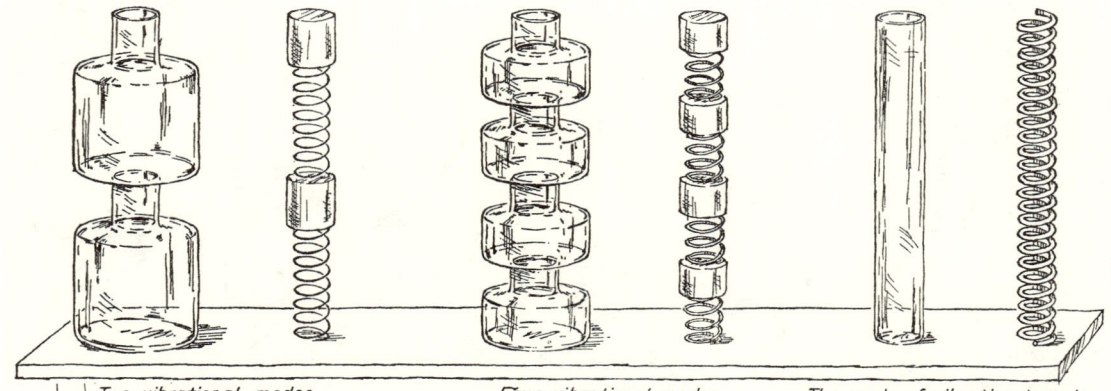

Two vibrational modes   Four vibrational modes   Thousands of vibrational modes

Fig. 30. Just as with balls on strings, the number of lumps in the bottle determines the number of vibrational modes. And as the string itself can be thought of as consisting of countless balls, so the cylinder at the right can be considered as composed of thousands of lumps and having thousands of vibrational modes.

Suppose now that we have made a set of odd-shaped bottles of the sort in Figure 30, and suppose also that someone has asked us to guess the nature of the vibrations that could occur within them. In the simple cases at hand (and without the confusing effects of more knowledge than we yet have at our disposal), it should seem quite obvious that mechanical cousins to these bottles can be made simply by replacing all the air in the necks by solid blocks, and all the volumes by weightless springs! This indeed is quite correct for the examples at hand; we could derive the modes of vibration of the bottles from a little experimenting with springs and masses. For example, the system consisting of two springs and two masses, which appears at the left of Figure 30, has two longitudinal modes of vibration: one in which both blocks move up and down in step with one another, and a higher frequency one in which they move in opposition. In Chapter III we saw diagrams of the transverse vibrations of string-mass systems, and it should not be too great a strain to see that the general behavior here is not very different, although much harder to draw. Suffice it to say then that lumpy bottles have their elementary modes of vibration, just as lumpy strings do, and that we can understand how a smooth pipe or horn behaves by extending our thoughts about bottles with many many lumps, in the same way we did for smooth and lumpy strings.

Experiments with Whistles

We cannot usually find or make the pretty beaded bottles I have been talking about, but we can experiment with many parts of the physics of bottles and horns using only odds and ends that lie around any house. For example, the effect of bottle volume on the frequency of vibration can easily be studied with the help of a pop bottle if you blow notes on it as it is gradually filled with water. With a little practice you can learn to excite the second, or even the third, mode of vibration of a medium-sized bottle. Because the tuning and tone color of wind instruments depend on the frequency ratios between their elementary modes of vibration, I want to emphasize to you that these relations are strongly affected by the *shape* of the air cavity in which the vibrations occur. Suppose I take an empty six-ounce Coca-Cola bottle and blow across it in such a way as to excite the lowest, and then the second lowest, frequency mode of vibration. With the help of a piano I find these two sounds have the pitches shown below, along with the calculated frequency ratio f_2/f_1 between these vibrations.

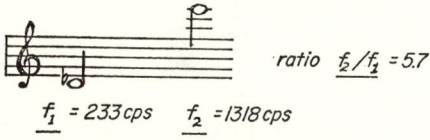

ratio $f_2/f_1 = 5.7$

$f_1 = 233\,cps$ $f_2 = 1318\,cps$

Suppose I now fill the bottle with water to a point just below the molded-in writing; the lowest mode will sound approximately F above middle C. It is a little difficult to coax out a clear second mode, but perseverance wins, and we find a new frequency ratio, as summarized below.

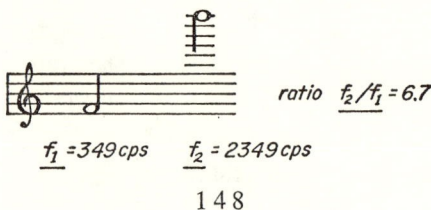

ratio $f_2/f_1 = 6.7$

$f_1 = 349\,cps$ $f_2 = 2349\,cps$

148

Pouring water into a bottle alters not only its volume but also its shape. In order to see what I mean, compare the cavity shape of an empty bottle with the extreme case of a bottle filled to within half an inch of the top. One cavity is "bottle-shaped," the other is almost cylindrical. You might find it useful to play around with some whistles in which only the shape is changed. The top diagram in Figure 31 shows how to make a tin or plastic cone. Try blowing across it at either end while closing off the other end with your thumb. Here the volume is left alone, but closing the large or the small end gives shapes that decrease or

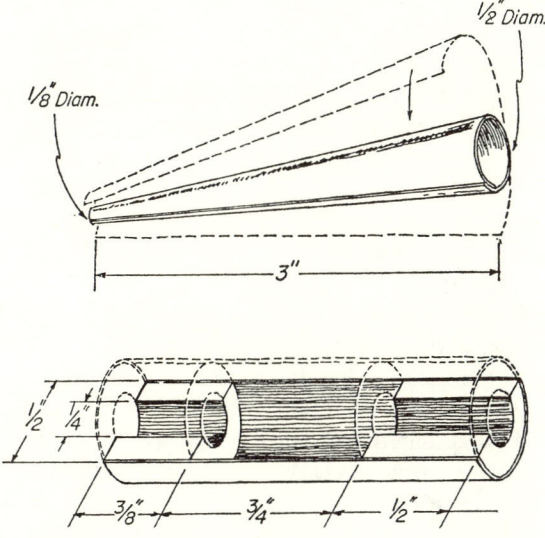

Fig. 31. A simple conical tube of tin or plastic (upper illustration) is fine for experiments on the relation between cavity shape and vibrational frequency ratios. The lower diagram is of Professor Sutton's trick whistle.

149

increase toward the open end respectively. My own whistle of this sort gives the notes and frequency ratios which follow:

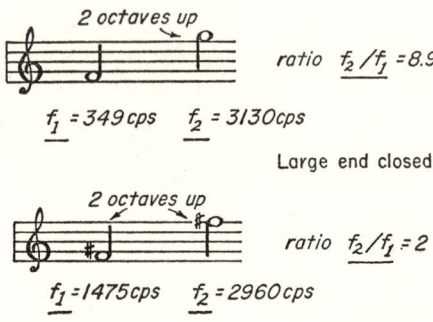

ratio $f_2/f_1 = 8.9$

$f_1 = 349\,cps$ $f_2 = 3130\,cps$

Large end closed

ratio $f_2/f_1 = 2$

$f_1 = 1475\,cps$ $f_2 = 2960\,cps$

Small end closed

You can see that there is quite a difference in the relations of the vibrational modes in the two cases, most of which you can explain on the basis of your present knowledge. A whistle with similar behavior can be made by cutting off the bottom end of a 30-caliber rifle cartridge, or by taking the rubber bulb off a kitchen basting syringe.

Professor Richard Sutton, who is well known for his sprightly demonstrations of physical principles, once showed me a whistle made like the one drawn in cross section in the lower half of Figure 31. Plastic is a good material for such a toy, preferably something opaque so that casual onlookers cannot see how it is made inside. Closing one end or the other gives a pair of whistles having very different pitches for the lowest two vibrational modes. The effect on a listener is quite startling, especially if you mix things up a little by making another one in which the narrow parts are of the same length. (Dr. Sutton may not forgive me for telling how the whistle is made, but perhaps I can rescue his favor by leaving its explanation to you.)

By now you are wondering why I have spent so much time on silly little noisemakers, and an explanation is indeed due you. Mathematical analysis tells us that the frequency of the lowest vibrational mode is determined *jointly* by the shape and volume of the cavity, while the frequency ratios f_2/f_1, f_3/f_1, etc., of the various modes to the lowest *are determined by the shape alone*. A great deal of time has been wasted by instrument makers and by laboratory workers in musical acoustics who are either ignorant of this theorem, or are unappreciative of its importance, which was first clearly recognized by Helmholtz. Since we base our understanding of *all* wind instruments on this point, I thought it would be helpful if I described little experiments first which might help you grasp some of its meaning.

Self-sustaining Vibrators

Among oboists and trumpet players, Johann Sebastian Bach's name brings thoughts of wonderful orchestral parts, marvelous in their music and also marvelously fatiguing to play. As some of you know, the Bach-playing oboist takes on a purplish tinge, while his colleague in the brass section turns just plain red. Obviously, these men are doing work—hard work— on expensive bits of plumbing that convert compressed air from their lungs into exquisitely shaped sounds. A part of our curiosity about musical instruments should be devoted to the simple-looking machinery at the mouthpiece end, which has the job of converting steady blowing into vibrations in the air.

Figure 32 shows one of our resonating cavities supplied at the closed end with a frictionless, leakless piston of metal mounted on a spring. If we were to set up one or another of the normal vibrational modes of the air in the cavity, there would be a simple har-

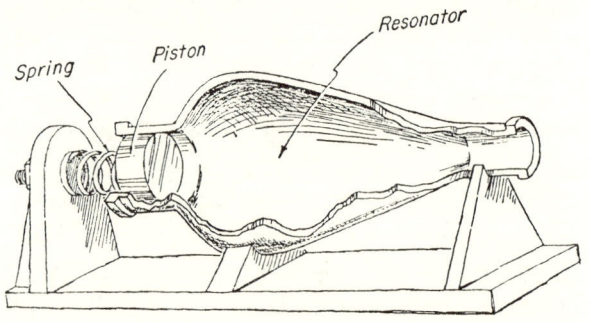

Fig. 32. Cavity resonator closed at the bottom end with a spring-mounted piston.

monic variation, above and below atmospheric pressure, of the air pressure acting on the inner face of the piston. This piston, in conjunction with its mounting spring, would think of itself as a simple driven oscillator of the sort we met in Chapter II. If the natural frequency of the piston and its spring is *higher* than that of the cavity oscillation which is trying to drive it, then you will recall that it moves back and forth in step with the pressure variations. By this I mean that the piston will find itself moving outward at those instants of time when the vibrational pressure on the piston is *above,* and moving inward when the cavity pressure is *below,* that of the outside atmosphere. If, on the other hand, the natural piston frequency is lower than that of the pressure variations, the piston will be moving in apparent opposition to the forces acting on it, as we learned earlier. Now we are able to inquire usefully about one of the principal requirements which must be met if a cavity is able to set up and sustain vibrations within itself with the help of a reed.

Suppose we drill a small hole in the side of the sleeve in which the piston moves, as shown in Figure 33, and suppose further that this hole is connected to a supply of air under pressure. If we can arrange that

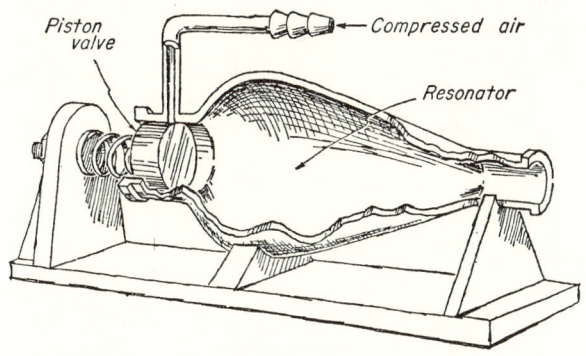

Fig. 33. A spring-mounted piston installed in this apparatus will act as a valve and excite vibrations in the cavity.

the motion of the piston is such as to admit new high-pressure air into the cavity at just the proper instant, we have a self-sustaining oscillator. A child's swing is a very similar sort of oscillator, in which the energy-replenishing pushes are timed to proper relation with the motion by means of someone's nervous system. A clock or watch is another example of such a "regenerative" system, as is the transmitter of your favorite radio broadcasting station. In this experiment (Figure 33) let us first decide what *is* the proper instant to inject the high-pressure air, and then it will be easy to see how to get the desired driving. If there were *no* source of new energy in operation, the vibration of the air in the cavity would gradually die out as its energy was dissipated in friction and in radiation of sound into the surrounding room. As a result, the amplitude of the pressure variations would die away in the manner illustrated in the upper line of Figure 34, which is a cousin of Figure 1, Chapter II. If we can raise the pressure amplitude after each full cycle by the amount it loses to friction and radiation, we shall have a steady oscillation of the sort shown in the second line. Here things are so arranged that the valve opens at the instant the cavity pressure reaches its maximum value, so that the external air supply can instantly "pump up" the bottom of the cavity, restarting the oscillation at its original pressure amplitude.

In wind instruments the players' lips, or the reed, serve as the vibrating piston-like valve, and must be able to restore 1 to 5 per cent of the amplitude in each cycle of oscillation. Flutes, tonettes, and bottles are excited in a different way, at the open end of the cavity. E. G. Richardson has made quite a study of such oscillators. You would be interested in one of his books, *The Acoustics of Orchestral Instruments,* E. Arnold and Co., London, 1929.

The second and third lines of Figure 34 together will help you to see that the valve must admit air to

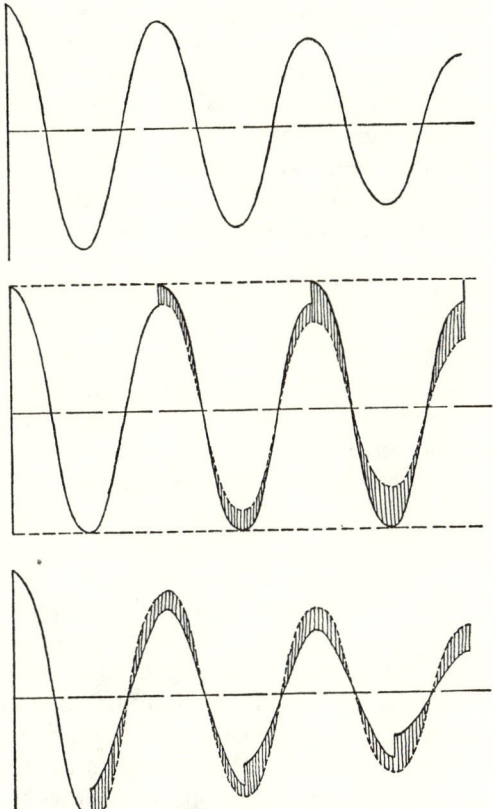

Fig. 34. *Top curve shows pressure variations at the bottom of a resonant cavity; the amplitude dies away from energy loss. Center curve shows how oscillation is sustained when fresh compressed air is admitted at proper time. Bottom curve shows oscillation killed when compressed air is admitted at the wrong time.*

the cavity during the peak of its own pressure variations (as in line two) if oscillation is to be maintained. On the other hand, letting in air at the time of a pressure minimum will serve only to kill off the oscillation in short order, as shown in the lowest line. If you are curious about such things, you will notice that these phase relations can be reversed if a suction pump is used instead of a compressor as the fundamental energy source.

All this shows us that a spring-mounted piston can serve as a valve for controlling a compressed air supply *only if its own natural vibration frequency is higher than the vibration frequency of the cavity which it is supposed to sustain.* It is only then that the piston motion has the right phase with respect to the vibrational pressures which drive it.

Some Complications

So far I have acted as though the whole business of making a cavity-controlled oscillator is a simple matter, to be explained in a few pages. As a matter of fact, we are really peering into a hornets' nest of complications, some of which have never been studied properly. Indeed, certain parts of this subject could not have been studied fruitfully until recent years because certain branches of vibration theory did not yet exist.

FIRST COMPLICATION

When I suggested that we imagine a piston mounted in the bottom of an air-filled cavity, I said nothing about its weight, and only made some remarks concerning its natural frequency of vibration. Such a simplification has at least let us learn something useful about the bare necessities of regenerative air-driven systems, but now we must return more nearly to the everyday world of practicality. You no doubt know, or can readily believe, that there are many ways to get a desired natural frequency from a spring-and-mass system. If the mass is large, we need a stiff spring, while a small mass can be attached to a relatively weak spring. Now, when we try to drive such a system at a frequency lower than its natural frequency, it turns out (as we might expect) that the weaker the spring is, the larger the amplitude which can be forced on it by a given magnitude of driving force. It is also true that under these conditions of below-resonance driving, the piston mass plays a small but important role. In our self-sustaining resonator system, the forces available for driving the valve-piston (or reed, as I shall call it from now on, whatever its real shape or material) are really very small, and the piston spring must be quite flexible if it is to be moved enough for the valve action to work.

Along with the spring, we need also to recall that the total driving force on the piston depends on its exposed area; a wide piston will be moved greater distances than will a narrow one of the same mass, if they are mounted in similar cavities. Thus we can say that the mass-per-unit area of the reed must be less than a certain amount if it is to respond properly to the instructions it gets from the cavity oscillations.

SECOND COMPLICATION

Closing up the bottom end of a cavity with an elastic reed has exactly the same effect on the natural vibrational frequencies of the cavity itself as does replacing the reed by a small neck and cavity with the same natural frequency and "effective mass" as the reed. The result of this is to alter the vibrational mode frequencies of the cavity, even when it is not being blown. Figure 35 shows what I am talking about; the upper part gives you an idea of what the reed-equivalent cavity might look like, while the lower part shows how this additional cavity alters the locations of the vibrational mode frequencies of the cavity resonator as a whole.

The practical implication of this complication, which was extensively studied by Bouasse, is that our musical wind instruments always play just a little *lower* in pitch when excited by a reed than you would expect from resonance measurements made on the air cavity with the reed clamped in its *closed* position. The flattening is generally a small fraction of a semitone in the lower registers of an instrument's musical range, but can become much more if high notes are played that are close to the reed's own natural frequency. There are, however, several conditions where such a correction is so much smaller than some others that it can go unnoticed.

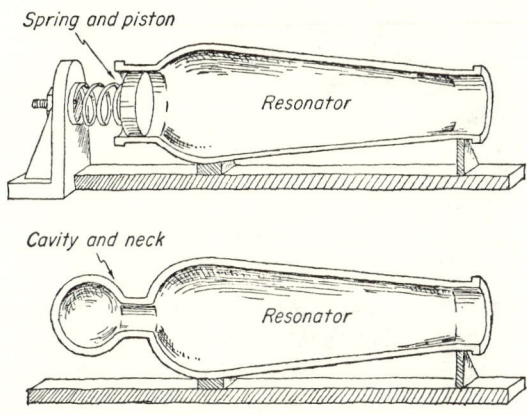

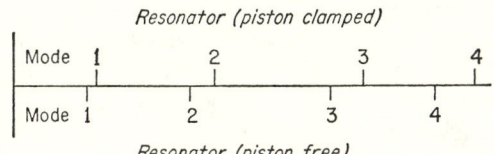

Fig. 35. Upper diagram shows the spring and piston system of Figs. 33 and 34 replaced with an equivalent cavity of the same natural frequency. If the natural frequency of the spring and piston system is higher *than any of the vibrational mode frequencies of the main cavity, the combined system vibrates (*lower diagram*) at slightly lower frequencies.*

THIRD COMPLICATION

The reed systems of most wind instruments are able to vibrate under the action of air pressure even when they are not attached to a musical instrument (oboe reeds quack beautifully when blown *sans hautbois*),

159

and we find to our interest (or horror) that the frequency of a blown reed depends strongly on the wind and lip pressure. Worse yet, it *never* emits a sound whose frequency matches that which the same reed calls its natural (unblown) frequency—the frequency you find by plucking at it in the manner of our old hacksaw blade. Our earlier investigations showed us that the natural frequency of the reed must be higher than that of the sound it is supposed to excite in the instrument, and here we have found that there are two kinds of reed frequency, blown and unblown. The aerodynamic forces on the reed (similar to those which make a flag flutter) contribute greatly to the change of frequency brought about by blowing. To make a long story short, I shall say merely that it is the *blown* reed frequency that counts, and prove it by telling you that the plucked-reed frequency of a clarinet is around 800 cps, while the highest note people ordinarily play is nearly 2100 cps, something which is manifestly impossible if the plucked frequency is the important one.

Let me summarize the complications we have met in this section: there are *three* frequencies we have to worry about in the theory and practice of wind instrumentology—the cavity resonance frequency, the unblown (plucked) reed frequency, and the frequency with which the reed vibrates when it is blown at a definite pressure. Even if you can do no more than keep this in your mind, the progress of musical science will be greatly helped!

FOURTH COMPLICATION

Up until now I have been careful to pretend that only one or another of the modes of vibration of the cavity has been kept in oscillation by the reed, when as a matter of fact all these modes are simultaneously excited to some extent. You will recall that I took some pains to show (in Chapter V) that the ear has the property of preferring sounds whose vibration recipes contain only frequencies which are integral multiples of the fundamental frequency. I used this fact to deduce the musical usefulness of uniform vibrating strings, since their modes of vibration possess the desired frequency. Now the time has come when I must tell you also that the vibrating air cavities chosen for wind instruments are those which have this property. Just what these shapes are can wait until the next pair of chapters—at the moment it is needful for you to know only that the musically useful vibrators are the ones whose higher mode frequencies f_2, f_3, f_4, etc., are two, three, four, etc., times the frequency f_1 of the lowest mode.

People who know their higher mathematics will tell you that a valve which lets puffs of air into a cavity at a definite frequency is in reality producing a complex driving sound with a vibration recipe whose ingredients have frequencies that are integral multiples of the puffing frequency. The amount of each ingredient depends on the way the valve opens, and therefore on the "shape" of the puff. The breaking down of puffs into their ingredients is called Fourier analysis by mathematicians. There is a close relation between this kind of analysis and our problem of finding vibration shapes of oscillators, but the connection is not as easy to explain as you might expect unless you are familiar with partial differential equations.

Figure 36 shows the recipes for three different "shapes" of puff: if the valve flies open and shut abruptly, admitting air for only about one twentieth of a cycle, the recipe contains approximately equal amounts of all harmonics, as shown in the first line of the diagram. The second line shows the recipe produced by a valve that opens abruptly for one fourth of a cycle and then closes sharply. On real instruments the valve opens more or less sharply, and for various fractions of the cycle, the precise manner being determined partly by the desire of the musician for one tone quality or another, and partly by the nature of the reed system. Because the harmonic frequency series of ingredients supplied by the reed matches the harmonic resonating frequencies of a *musical* horn, we find that *all* the vibrational modes of the horn are excited by the reed, to an extent which depends on the damping of each mode. Not only are all the vibrational modes of the horn driven by air from the reed, but all these vibrations react back on the reed itself in a very complex way. The result is that the vibration recipe of the air puffs put out by the reed depends in an additional way on the shape of the horn. As a matter of fact, this dominance of the horn on the reed system is often so strong that the various band instruments keep a great deal of their characteristic tone even when played with a reed belonging to another wind family. Later on I shall give you specific examples.

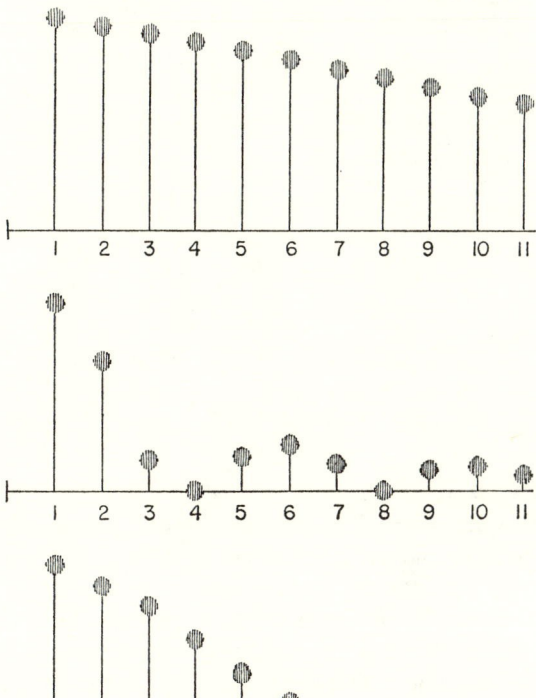

Fig. 36. *The vibration recipe of air puffs from a reed valve depends on the shape of the puffs. Top: A puff of air lasting* 1/20 *of a cycle has almost equal amounts of the first dozen or so ingredients. Middle: A series of puffs, each lasting* 1/4 *of a cycle, has these ingredients. Notice that the fourth, eighth, etc., ingredients are missing; there is a close similarity between these phenomena and those illustrated in Fig. 14. Bottom: This recipe belongs to puffs lasting* 1/8 *of a cycle.*

163

CHAPTER VIII

The "Brass" Instruments

People were still hiding in caves and swinging in trees when they discovered that it was possible to make fierce and terrifying sounds by blowing through pressed lips into the opened small end of a hollow animal horn. Thereby was founded the trade of trumpet player. Jewish religious ceremonies still make use of the ram's-horn shofar; in India the conch shell is used for the same purposes, and I have heard soul-stirring trumpets in church on Easter Sunday. Not only are such "lip-reed" instruments closely connected with most of the world's religions, but they also have an important part to play in the business of war. It is commonly known that a well-played bugle or trumpet makes almost anyone feel brave and warlike, but very few generals have been able to emulate the example of Joshua in toppling the walls of Jericho with blasts from a corps of trumpeters. However, my interest in horns, and probably yours, is placed in the more peaceful concert hall and football field.

The general name "horn" for the modern group of

lip-reed instruments has an obvious historical origin, while the alternative name "brass" (or "copper," as the French would have it) comes directly from the materials used for their manufacture. Let me begin by describing the family briefly. A brass instrument starts out with a small cup-shaped mouthpiece that is coupled to a long pipe, which ends in a flaring part called the bell. Instrument makers like to call the whole inner volume of a wind instrument the "bore," and since this is a short and convenient catchword, I shall use it from now on. In the brass instruments the air in the bore can be set into vibration at one or another of a certain set of frequencies, with the help of the regenerative action of the air-pressure variations on the player's vibrating lips. Things are not quite as straightforward, however, as my discussions of the last chapter would imply. If you could snatch the bugle away from a bugler while he played a note, leaving him his mouthpiece undisturbed, you would hear the buzzing of his lips at a pitch that is close to his previous one, although the tone quality and loudness would be greatly altered. You will understand what I mean if you ask a trumpet- (or better, French horn-) playing friend to sound tunes for you on his mouthpiece alone. Not only can a brass player give you the desired notes without his instrument (albeit shakily); upon request he can play his instrument a little bit sharp, or an almost unbelievable amount flat, in pitch, an ability that is, of course, a very important part of the technique of competent jazz musicians. It is obvious from this that a lip reed is not at all in the mood to submit meekly to the dictates of a coil of brass tubing, even though it pays *some* heed.

We must remember that the brass player "makes" the note, and his lips can do some very complicated things to the physics of musical brasses, not all of them understood scientifically.

Let us see what is the cause of our lips' disobedience to the horn, and find out what new things it can hint to us. Our lips are rather heavy, so that the air within a bugle cannot shake them easily and thus determine their frequency of vibration. The vibrating air can, however (and this is what makes brass instruments practical), give strong "suggestions" to our lips, so that it is easier to make them go at one or another of the favorite frequencies of the horn. Because of the "suggestions," a player has only to get his lip tension reasonably correct for the note he wants, and allow the instrument to "pull it in" near the proper pitch. Conversely, if the instrument is not itself perfectly in tune, the player can still keep his job by "lipping" the note up or down to the correct place.

Privileged Frequencies

Let me describe to you a simple experiment that I did one afternoon recently with a piece of iron pipe whose inside diameter (I.D.) is about three quarters of an inch, and whose length is very nearly thirty-three inches. Any pipe of approximately this diameter will work quite well. My favorite material for such informal experiments is the steel tubing sold by electrical contractors and hardware stores under the name "thinwall conduit." This comes in half- and three-quarter-inch sizes, which are perfect for acoustical experiments, and costs only about $1.50 for ten feet.

Vibration theory tells us that such a cylindrical pipe has a lowest mode frequency close to 100 cps when one end is closed off, and the second, third, and fourth modes have frequencies that are about 300, 500, and 700 cps, respectively. When I put such a pipe to my mouth with the idea of playing it like a bugle, my lips form a closure at one end which pretty well satisfies the assumptions of my calculation and also provides a reed system with which to excite the vibrations of the pipe. Now, if we take our simple ideas seriously of how reeds conspire with cavities to sustain an oscillation, we should expect that my lips would like to vibrate at 100, 300, 500, or 700 cps. Let us see now what actually happens. Please bear in mind that since my experience in playing brass instruments totals about thirty minutes spread thinly over the last twenty years, my ability to coax out a difficult note is limited. The lowest note I could get, and it took a little effort, had a frequency quite close to 100 cps. This I expected. As I gradually tightened my lip tension, the pipe made recognizable attempts to dictate to my lips near each of the following frequencies: 125, 150, 165, 175, 225, 233, 250, and 300 cps. The last of these frequencies gave the pipe quite a strong control over the reed system, as indeed we might expect, 300 cps being one of the vibrational frequencies beloved of the pipe. I did not go to higher frequencies for two reasons, one of which overshadows the other: there was nothing essentially new to be learned from the higher notes, and my inexperienced lips were just not able to muster enough tension to reach these higher notes without a smaller mouthpiece.

Before I discuss the unexpected new notes in this series, it is probably worth a moment to confess that the numbers I have quoted are somewhat rounded off and smoothed to make the explanation easier. They were all obtained by comparing my sounds with various notes on the piano, after which a table of notes and their frequencies was used to find my numerical results. I should also say that if you want to make the experiment more comfortable for yourself by using a regular mouthpiece, be sure it is one taken from a trombone, baritone, or tuba. A trumpet or bugle mouthpiece is designed to work at much higher frequencies, and is unlikely to permit you to get down into the two lowest modes of a thirty-three-inch pipe.

If you look closely at the sequence of frequencies I have obtained, you will suspect the presence of some whole-number relations among them, and this is indeed a correct suspicion. Bouasse showed by many different kinds of experiment that a lip reed or other reed *too massive to be dominated* by the air in the instrument bore not only will prefer to vibrate near one or another of the normal mode frequencies of the bore, but finds itself amenable to suggestions at a set of what he called "privileged frequencies." These privileged frequencies are found at integral *sub-multiples* of the pipe's own vibrational mode frequencies. That is, an instrument which has one of its natural frequencies at 600 cps will show privileged frequencies at 600/2, 600/3, 600/4, etc. cps, along with similar descendants of all the other vibrational modes. You will find an interesting set of privileged notes on a trumpet if you can attach a trombone or baritone mouthpiece to it. If you are already a player of brass instruments, your trained lip may make it a little hard to find these privileged notes by going *up* the sequence from the lowest one. Start with the highest one and search for the lower ones by gradually slacking your lip tension. The pitch will then "hang up" on each privileged note as you pass it.

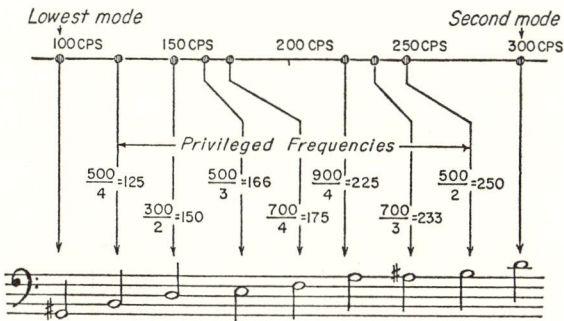

Fig. 37. A piece of conduit played as a bugle has these favored frequencies. The lowest mode has a frequency of about 100 cps; the next one is found near 300 cps. All the in-between frequencies are privileged. Their ancestries are shown in the middle of the diagram. The musical staff gives the approximate pitches of all these notes.

With the privileged sub-multiples in mind, it is not very hard to disentangle my pipe-blowing results, which are given diagrammatically in Figure 37. The top line shows the frequencies that were chosen for preference jointly by the pipe and my lips, while the labels above and below this line tell how each of these favored frequencies came about. The second line shows *very* approximately the notes on the piano nearest to these frequencies, as a help to you in recognizing them if you should care to repeat the experiment. I cannot explain the origin of these notes in a non-mathematical book, but it may help you a little to know that they are distant cousins of the combination tones we met in Chapter IV. Having thus put you off as to the explanation of privileged frequencies, I must at least tell you that regardless of their cause, they will have an important bearing on the sort of bores we can use in building musical instruments.

Several years ago I gave a little thought to how a reed system can maintain oscillations in the air of a musical instrument bore, and found mathematical hints that phenomena similar to the privileged frequencies might very well exist. It was a real pleasure later on to run across Bouasse's description of his work. He does not come right out anywhere and give a proper explanation of the effect, but tucked away in an appendix of one of his loosely organized books on musical physics is an analysis of a certain unusual type of oscillator whose properties are closely related to musical reed systems. In 1929 there were not many people who even knew of this particular oscillator, so Bouasse's interest in it makes me think that he knew the explanation of his privileged sounds.

An Especially Simple Horn

Bugles are the simplest of modern brass instruments, and as such can provide us with a stepping-stone into the realm of orchestral horns. At first we can pretend that the bugle has a simple conical bore (which is coiled up only for convenience in carrying it), closed at the small end by our lips applied to a cup-shaped mouthpiece. Such a conical bore closed at the small end can be shown mathematically to possess a complete set of harmonic vibrational frequencies, so that the second, third, and fourth modes have frequencies that are double, triple, and quadruple, respectively, that of the lowest mode. I am describing here what is sometimes called a "cavalry bugle." The everyday bugle one sees is not conical, and is more correctly called by its army name, "field trumpet." This last instrument is, however, quite similar to our cone in its behavior. Strictly speaking, for the cone to have *precisely* harmonic frequency ratios between its modes of vibration, the cone should be complete; that is, it should extend all the way to its point. It is a curious (though mathematically straightforward) fact that cutting off any amount of the small end less than about a third of the length of the complete cone will give a system whose vibrational frequencies are almost identical with those of a complete cone if the small end is closed. You can get some idea of the plausibility of this if you consider that the small volume of air contained in the tip of a cone constitutes a rather stiff spring, which can be replaced by a perfectly rigid closure with little effect on the vibration of the remaining air.

Suppose we play on our simplified bugle now, and find some of the more interesting frequencies it will produce. On the assumption that its complete cone would be about sixty-six inches long, its lowest vibrational mode has a frequency near 100 cps (a cone has to be twice as long as a cylindrical pipe if their lowest modes are to coincide in frequency). We expect that the vibrational modes of the cone will give strong admonitions to our lips, but what about the privileged notes? Dividing the vibrational mode frequencies by various integers will give us an array of new notes as before, but certain of these show a curious relationship. Frequencies which the cone considers its own vibrational property turn out also to be privileged, and several times over, at that! Let me illustrate my meaning with a short table:

Cone frequencies:	100	200	300	400
Matching privileged frequencies:	200/2	400/2	600/2	800/2
	300/3	600/3	900/3	etc.
	400/4	800/4	etc.	
	500/5	etc.		
	etc.			

We are led to see from this that it should be quite easy to play all members of the cone's family of harmonics since many things work together to give the cone dominance over our lips. These harmonics are indeed so much more easily excited on a cavalry bugle that most players are completely unaware of the existence of the more motley members of the privileged tribe of sound. As a result a military bugler can skip nimbly back and forth between vibrational modes of the cone as he awakens his colleagues or tells them to go to bed.

Music for bugles is usually written so that the pitch of the lowest vibrational mode of the instrument appears as the C below middle C, and as a result the next few modes have written pitches corresponding to middle C, G, C, E, and G (in ascending order). Most buglers would find it hard to blow the low C on their instruments, which they call the "pedal note"

172

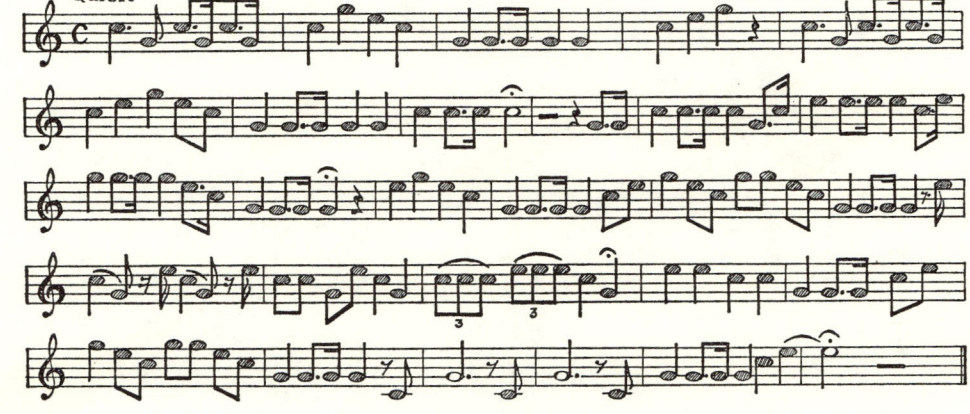

Bugle call: "Retreat"

or "fundamental," but this causes no trouble since composers have found almost no use for it. The use of the word "fundamental" in this connection appears proper since it refers to the lowest member of a true harmonic series, and also to the frequency of the lowest mode. However, we shall soon see that this is a rather exceptional case. The "pedal note" on most modern brass instruments is *not* usually associated with the lowest mode of the horn!

On page 173 is a rather beautiful example of bugle music, which may be familiar to some of you. Notice that the pedal note is never called for, and that its octave comes in only seldom. It takes a great deal of ingenuity to create real music using a total of only five notes, but it obviously can be done, and now and then you will hear even ensemble playing with bugles with quite striking harmonies. (You will notice that a set of four bugles cannot possibly play a "wrong" chord in the simple-minded sense of Chapter V.)

Tuning Problems and a Fairy Tale

If you compare the shapes of the different varieties of orchestral brass instruments, you will find none of them exactly conical in shape, and most of them are even cylindrical for a considerable part of their length (the trombone is a prime example). You have learned that a uniform pipe has its modal frequencies spaced out at 1, 3, 5, 7, etc., times that of its lowest mode, and that a conical horn has its frequencies spaced out at 1, 2, 3, 4, etc., times the lowest mode frequency (which is an octave higher than that of a pipe of the same length). Since the shapes of real brass instruments are intermediate between these two extremes, you are entitled to deduce that their normal mode relations will also be intermediate. That is, a partially cylindrical, partially tapering horn (which Bouasse aptly calls "cylindro-conical") will have its lowest mode somewhat higher than that of a uniform pipe of the same length, and its next mode is between two and three times higher in frequency than its first mode, and so on. Such a horn seems to be useless because it has abandoned the two types of bore which give the exactly integral frequency ratios we have so often found to be necessary in music. Things are not so bad as they seem, however, as we shall see with the help of a fairy tale inspired by some pages in Bouasse.

Let us first set the stage by looking at the following table, which compares the normal mode frequencies of a thirty-three-inch pipe with those of a conical horn having the same length, their lowest modes vibrating at about 100 and 200 cps, respectively. The lowest line in the table gives the musical interval in semitones between the sounds of the two bores when they are excited *in the same mode*. You will remember that a semitone is the pitch interval between adjacent notes on the piano, or between successive notes in a chromatic scale. The frequency changes by very nearly 6 per cent when the pitch changes by one semitone. A whole tone is an interval of two semitones—for example, C to D.

Mode Number	1	2	3	4	5	6	7	8
Frequency of pipe	100	300	500	700	900	1100	1300	1500
Frequency of cone	200	400	600	800	1000	1200	1400	1600
Musical interval (semitones)	12.00	4.95	3.15	2.31	1.83	1.52	1.28	1.12

As you look at this table, you will notice along with
Bouasse that at mode number 4 the pitch difference
between the two instruments is only about one whole
tone, an amount some players might be able to pull
a note with their lips, while the discrepancy is down
to a semitone at mode 8, and decreases further in the
higher modes. Let me concoct the fairy tale to help
you to see the implications of this table.

Once upon a time a cruel witch decreed that a certain bugler must play on a uniform pipe instead of his
accustomed bugle, and do it in tune with other musicians playing ordinary instruments. Diligent practice
taught him how to "lip" his top two notes into tune,
but the lower three notes of his accustomed five-note
range were hopelessly out of tune. Now, this musician
was a clever fellow who realized that his tuning troubles were all one-sided—*all his notes were flat when
uncorrected.* A little thought suggested to him that
shortening his pipe a suitably chosen amount could
bring the lowest note up into correcting range while
making the highest one sharp in pitch, but still cor-
rectible downward. He realized that sawing off enough
to raise the pitch an amount equal to the average
error of his highest and lowest commonly used notes
might give him what he wanted, as shown in the next
table.

Mode number	1	2	3	4	5	6
Desired note	C	C	G	C	E	G
Discrepancy (semitones)	8.77	1.72	−0.08	−0.92	−1.40	−1.72

We can see that all the notes he has to play are in
error by an amount less than the two semitones
which is the range over which he can correct things

with his lip! It is also quite clear that the lowest mode
of his sawed-off pipe is still far from being useful to
him, so that over the years he forgets about it. One
reason he forgets easily is that he soon discovers a
good privileged note about where that pedal C ought
to be. He decides to call it the "fundamental" of his
horn and teaches his students to follow his ways!

My little fairy story is exaggerated, but like so many
stories, it has a moral for us. Any man who has a
good ear and a mind that works well enough to coax
a straight pipe to sound in tune with a bugle of nearly
the same length certainly will have no trouble in doing
much better with a cylindro-conical bore whose properties are not nearly so far away from those of the
well-tuned cone. His job is made easier for him by
the existence of the exactly harmonic privileged notes
which will be found clustered around each of the
slightly bent vibratory frequencies belonging to the
bore, so that they help his lip pull the notes into
tune.[1]

Lack of an ordinary vibrational mode at the pedal
note gives no one any trouble, as our imaginary
bugler found; the 1/2-frequency privileged note of
mode 2 coincides with the 1/3-frequency note of
mode 3 and so on to give a multiply defined note at
the pedal frequency!

[1] The vibrational frequencies of the air in a trumpet which
two of my students recently measured were such as to agree
very well with the desired harmonic series of notes, except
for the lowest mode, which was about five semitones flatter
than the pedal note.

Getting a Chromatic Scale: the Slide

While people have managed to write interesting music for simple brass instruments of the sort we have been talking about, the gaps between the available pitches are large enough to pretty well cripple a musician who wants to play ordinary music using the lower few modes of his horn. He can sometimes play such music up high, since his notes form a complete though approximate scale above the eighth harmonic. Lower down, he may be able to use some of the privileged notes as a crutch, but all in all the scale is pretty shaky. A hundred and fifty or two hundred years ago, when everyone played on such instruments, musicians had to specialize on one or another range of pitch, so that several men would take turns in the brass parts of a major composition.

A slide similar to the one used on a modern trombone was one of the earliest devices invented to help musicians escape from the normal mode frequencies of a simple horn. Because of its straightforwardness, we can use it to learn how all the brass instruments get their complete scales. While the slide is not marked in any way, every trombonist has drilled into him seven different slide "positions," each of which gives him a horn of a certain length, having a pedal note and a family of near harmonics. By convention, the first position is taken to be the one in which the slide is all the way in, so that the instrument is as short as it can possibly be. The second position has the slide far enough out to lower the pitch of any given mode one semitone, as compared with the first position.

The other positions lower the pitch a semitone at a time, to a total of six semitones. To keep things looking familiar, Figure 38 is arranged to show the behavior of a slide brass instrument of such a size that it will play along in unison with our conical bugle when kept in the first slide position. (This diagram

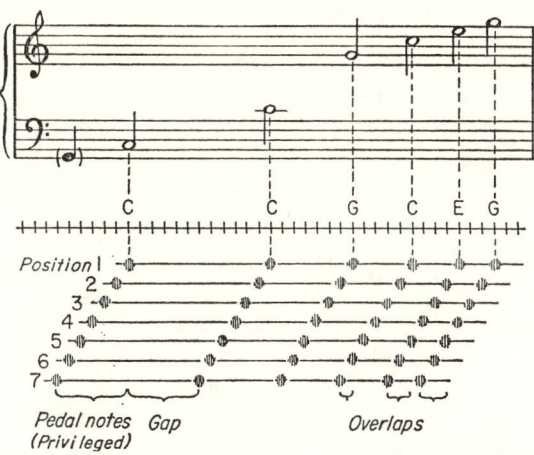

Fig. 38. The scale of a brass instrument is shown on the musical staff. The white notes are the musically used members of the harmonic series. The black note gives the position of a typical lowest vibrational mode. Running the slide out through the successive "positions" depresses the pitch of the whole instrument a semitone at a time.

will also turn out to apply directly to the ordinary trumpet and most other brass instruments, when we learn how to reinterpret the meaning of the slide position.) The "white" notes on the musical staff show the pedal note and its harmonics, which formerly belonged to the bugle, while the "black" note to the left gives you an indication of the actual pitch of a typical instrument's lowest vibrational mode. The row of shaded dots on the line marked "position 1" shows on a linear semitone scale the pitches of these same harmonics. Extending the slide to the second position lowers all the playable notes by a semitone, as shown by the dots on the next lower line, and so on through all the positions.

Notice that the first three positions serve to close the gap between the highest E and G belonging to the bugle's scale, while four positions are needed to give a chromatic scale from the C below this. The set of seven positions gives us a complete chromatic scale from the F below middle C all the way to the top bugle G. There is an unfilled gap between the highest pedal note (position 1) and the lowest note (position 7) of the second harmonic family, but this causes no trouble, since, as I remarked earlier, the pedal notes are not ordinarily used. Certain bass instruments do have provision for crossing this gap. We also see that there are overlapping ways to get some of the higher notes. For example, there is a choice for the C that lies an octave above middle C: it can be played as the fourth harmonic of the position 1 pedal note, or as the fifth harmonic of the pedal note for position 5. Brass instruments can be played in higher modes than my diagram indicates, and most can be coaxed up above the tenth or twelfth harmonic, but the principles involved remain unchanged.

Valved Instrument Scales

As you well know, the trombone is the only band instrument that has a simple slide with seven positions. The other brasses have a set of "piston valves" of the general sort shown in Figure 39, which, when depressed, add a definite length of tubing to the instrument bore. Usually only three such valve assemblies are used: one that adds enough tubing to lower the

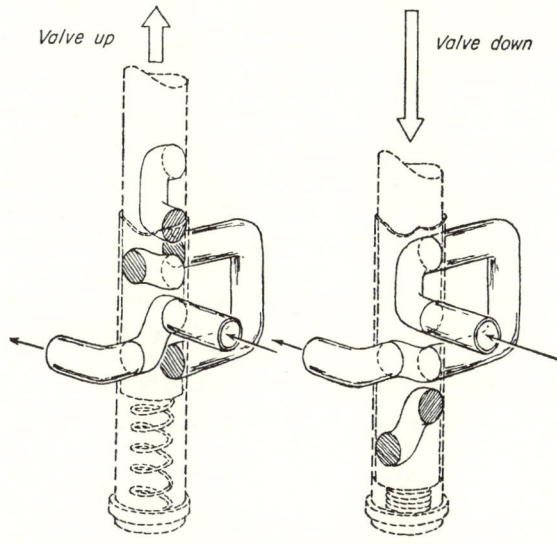

Fig. 39. How a typical valve for a brass instrument works. When the valve is in the up, or "open," position (left), there is no addition of tubing to the bore length. Depressing the valve piston (right) rearranges the air passages to increase the available length of tubing.

181

pitch a semitone, one that lowers it two semitones, and a third that depresses the note three semitones. When the valves are all "up," we have the equivalent of trombone position 1; pressing one or another of the valves gives the next three positions. At first thought, it seems obvious that the equivalent to position 5 (four semitones down) is obtained by pressing both the semitone and tone-and-a-half valves, and that position 7 (six semitones' depression) is obtained by pressing down all three valves. Let us find out what happens in real life, with the help of a simplified example I have worked out.

Consider a trumpet that is accurately in tune on all its "open notes" (valves in the "up" position) from middle C on up, but whose lowest vibrational mode is five semitones flatter than the desired pedal note, so that the pedal must be played as a privileged note. Calculation shows that if the three valves are to depress the pitch the exact amounts required when playing E, F, and F♯ above middle C (or any of the higher modes of these tube lengths), the amount of extra tubing added by each valve must be that given in the following table:

VALVE	LENGTH ADDED
1/2 tone	3.48 inches
2/2 tone	7.17 ″
3/2 tone	11.10 ″
Total addition	21.75 inches
6/2 tones	24.22 inches
Discrepancy (inches)	2.47 inches *too short*
Discrepancy (semitones)	.50 semitones *sharp*

Pressing down all three valves then will add the sum of these tube lengths to the bore, as shown. I have also computed the amount of elongation required to give an accurate lowering of six semitones in pitch, along with the discrepancy in inches of tube length and in semitones of pitch between the "correct" amount and what we get from three valves. These pitch and length discrepancies are calculated for notes from middle C on up; because of the flat first mode of the bore, the notes obtained by using these valves for notes below middle C will be *flat,* as follows:

VALVE	FLATNESS OF NOTES BELOW MIDDLE C		
1/2 tone	.21	semitone flat	
2/2 tone	.40	"	"
3/2 tone	.56	"	"
All three pressed	.97	"	"

A serious musician would be most dissatisfied with such an instrument, which would tax his powers of pitch correction to the limit. I have described all this to show you that we are faced with a problem which has *no* exact solution. Fortunately it *does* have several surprisingly accurate approximate solutions that can be chosen in the spirit with which our imaginary bugler worked, and in very much the same way as the even-tempered scale itself was chosen. Different manufacturers of brass instruments attack this problem in different ways and with different degrees of success. A proper discussion of their solutions would require many pages of dull and difficult argument.

You will notice that throughout this whole chapter I have never really told you *how* to calculate any lengths or other dimensions of brass instruments. This is because accurate calculations are not usually easy, and are often impossible. Yet you will often read statements to the effect that "brass instruments act like cylindrical pipes open at both ends" or similar statements about cones that are either open or closed at the small end, and there are great divagations on the harmonics, modes, and partials of simple pipes which have over the years covered up a great deal of the facts of life in the theory of brass instruments. People have spent hours trying to find out the "equivalent length" of the supposed simple pipe for some instrument, and all they are really doing is a job like that of our bugler who found a closed pipe that was workably but painfully equivalent to an ordinary bugle. No wonder musicians sneer at the follies of "scientists," and claim immunity from the laws of physics! Bouasse constantly snipes at this sort of naïve "research," and has written some positively sulfurous passages about people who take it too seriously.

We should remember constantly that brass instruments are neither conical nor cylindrical, and therefore their properties must be studied directly, either with experiment or with mathematics, and preferably with both. To a very good approximation, a brass instrument is closed at the mouthpiece end—closed, if you like, by a complex elastic system subject to aerodynamic forces, but still essentially closed. If you want to calculate lengths for valves, or for the whole bore for that matter, you should solve the proper equation for that particular shape of bore, being grateful for any mathematical simplicities you may find, but expecting none. The final test is that of the musician. Can he play your instrument in tune? Does he find it responsive? Can he control its tone and loudness? You needn't always take his verbally expressed reasoning too seriously if it concerns physics, but he is almost always right in what he says happens, or in what he says he wants. I can assure you from personal experience that it is an enlightening experience for a physicist to find some of his little mental pets cross-breeding with those of practicing musicians; the hybrids are sometimes more useful and vigorous than elegant, as is also the case with mules!

What Determines Tone Color?

If you attach a mouthpiece to a cylindrical pipe, you can play a set of muffled and unclear sounds, whose frequencies we have already discussed. What about the tone color and the vibration recipe of these sounds? What is the relation of these rather unpleasant noises to the noble blare of a true brass instrument? Part of the answer lies in the effect of the horn on the "shape" of the puffs of air which are made by our lips as they vibrate in the mouthpiece (see Chapter VII), and part is descended from something I told you back in Chapter II. While we were experimenting with a hacksaw blade, we noticed that rapid vibrations were more easily heard than were slow ones, and I explained that part of this came from the changing ability of a vibrator to produce sounds in the air. In a wind instrument the vibrating object of interest is the air in the bell, and it is not hard to imagine that changing the bell shape can affect the relation between the vibration recipe of the air in the horn and that of the sound we hear. One is directly caused by the other, but they are not identical.

Because it is a dimension that will come in frequently in the next few paragraphs, I should remind you that the distance sound travels in air in the time of one vibration is customarily called the wavelength of the sound. It is a perfectly definite length, which is greater for low- than it is for high-frequency sounds, and the dimensions of our horns relative to this length will determine their sound radiation properties.

Not only does an aperture large or small compared with the wavelength constitute a good or poor radiator of sound, but the effective aperture size itself of a flaring horn-shaped bore depends on the rate of flare. A horn whose cross-sectional area increases only little by little for distances of one wavelength at a time behaves as though the radiating aperture were pretty much the size of the large open end of the horn. Conversely, a bore whose cross section changes drastically in the distance of one wavelength radiates as though it were an aperture about the size of the *small* end!

Let us imagine that we have smuggled a tiny microphone into the bore of a certain brass instrument, and that analysis of the pressure variations detected by this microphone shows that the vibration recipe *inside* the bore is something like that shown in the top part of Figure 40. Let us also imagine that we have somehow measured the sound-radiating ability of our instrument, and have found that the bell acts like a small aperture for sounds with frequencies up to that of the fourth harmonic of our played note, but considers itself a nearly perfect radiator for all frequencies higher than this, as indicated by the middle part of Figure 40. This is because the ordinary bell makes its whole expansion in a distance that is less than a wavelength of the lower components. If we put these two oversimplified and imaginary sets of experimental data together, we might guess that out in the open air we

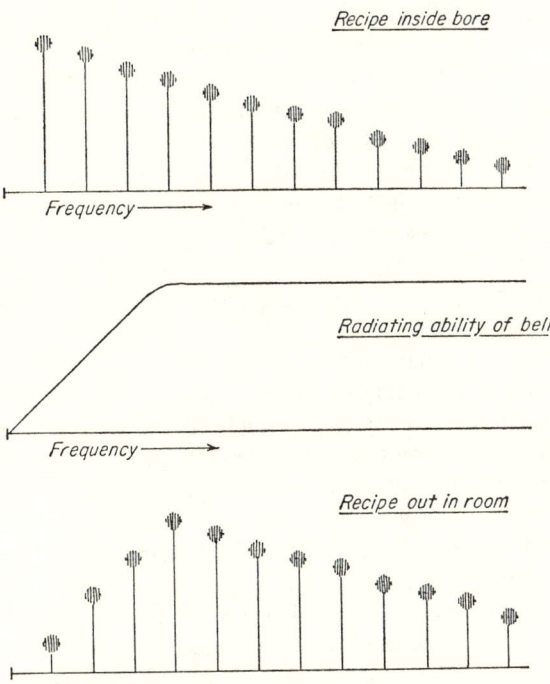

Fig. 40. Because of the radiating properties of the bell of a horn, the sound we hear has a vibrational recipe different from the recipe our lips set up inside the bore of the instrument.

would find the bell discriminating against the lower few components of the vibration recipe, while the higher frequency ingredients are pretty much the same as they were inside the bore, in some such way as that shown in the bottom part of Figure 40. I must emphasize that for real instruments the diagrams corresponding to those in Figure 40 are much more

187

complicated in their shapes, and in their relative proportions, but the over-all behavior is rather similar.

The particular frequency at which the radiation becomes really efficient depends on the shape of the bore, and especially on the form of the bell. This is a dominant reason why different members of the brass choir have different tone colors; even if identical mouthpieces were used (so as to get similar recipes *within* the bores), the different bell shapes will have different ways in which they emphasize the various vibration ingredients.

The familiar mute gives a good example of the effect of alterations of the shape of an instrument bell on the tone quality of its sounds. The mute also calls to our attention the interesting fact that it is possible to drastically alter the radiation properties of a bell without upsetting its tuning! We find that the tuning depends very heavily on the shape of the bore in the first three quarters or so of the over-all length, while the tone quality chiefly depends on the last third, if we leave aside the behavior of the mouthpiece. You can verify this for yourself with the aid of iron pipes of different lengths used with and without a funnel or homemade tin-cone "bell." Any mouthpieces you can borrow will add to the variety of your researches. The characteristic "brass" sound is quite clearly dependent on the presence of a bell of some sort. This is a fortunate circumstance for the already hard-pressed designer, because it gives him room to maneuver as he threads his way among a forest of conflicting requirements.

Real Brass Instruments

We have been wandering around peering at brass instruments now for most of a chapter without really getting close to them. We are somewhat in the position of a biologist who has raised mice in a laboratory and who has read books on other kinds of animals but who has never had a chance to go out in the woods to watch wild animals at their work and play. In order to change this situation a little, I should like to call your attention to the photographs of a trumpet, trombone, and French horn, which appear as Plates II, III, and IV. The loops of tubing that are introduced by the three valves are clearly visible in Plates II and IV; the middle valve in both adds the short (semitone) loop, while the valves nearest and farthest away from the mouthpiece add in the tone and tone-and-a-half loops, respectively.

Several times in passing I have remarked on the effect of the mouthpiece itself on the tone quality of an instrument, and it is time here to say a little more about it. In Figure 41 you can see cross-section drawings of the bores within mouthpieces that are used with our three instruments. It is an experimental fact that the depth of the cup affects the solidity with which a player can "lock in" on a desired pitch, as well as affecting the tone color of the instrument. Whether the bottom of the cup has a sharp edge as in the trumpet mouthpiece, or has none as in that of the horn, is important. The effect is usually described thus: a sharp-edged cup gives a "hard," "incisive" tone, while rounding the edge off, or removing it altogether, gives a "softer," "smoother" tone. This is

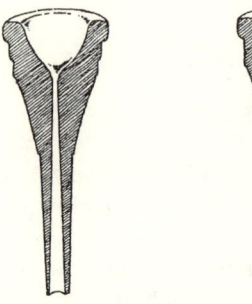

Fig. 41. These are the two extreme patterns of brass instrument mouthpiece bores. The cup-shaped one at the left is typical of the trumpet and trombone mouthpiece. The pattern at the right is for a French horn. Other brass instruments use a shape intermediate between the two.

perfectly true, but these words are in musician language, and are therefore essentially metaphorical. I have not heard a really thoroughgoing explanation of the effect based on fundamental principles of physics, and no doubt the explanation will prove rather difficult and subtle. However, no one who says, "Yes, of course, sharp corners would naturally give sharp and harsh sounds," is going to ask himself how to do the physics. He is in the same intellectual hole as the characters who believe that eating meat is the way to become brave, because fierce lions are carnivores, while peaceful cattle are vegetarians. How do these people explain the dangers faced by a bullfighter, or the comfort they themselves take from the loving attention of the family pussycat?

Be that as it may, players on the trumpet and on the trombone can be quite confident of hitting the desired harmonic, and they produce an incisive tone, when using their own traditional mouthpiece. On the other hand, I know of one major orchestra in which the French-horn section once had a standing agreement: they would go out and drink champagne after any public performance in which none of them burbled or otherwise missed a note. Sad to say, these men never got their fill of champagne! Not only does the horn player have to contend with a "shoulderless" mouthpiece, but his job requires him to play constantly in the very high harmonics of his pedal note, where things have all run together.

It is a meaningless task to give you a detailed account of the dimensions of our three instruments, since they differ from manufacturer to manufacturer. I can give you, however, a summary of three particular instruments, along with a cross-section diagram of a particular trumpet (Figure 42) drawn with its transverse dimensions five times too large, to show the details better. You will see in the tabulated summary below the percentage of the total length of each

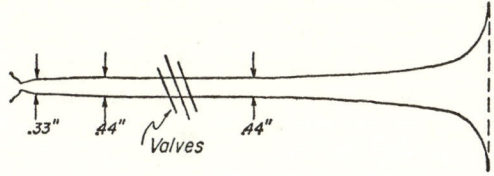

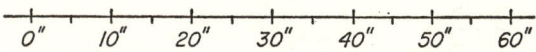

Fig. 42. Cross-section diagram of a straightened-out trumpet. Diameters are drawn to a larger scale than the lengths; the instrument looks five times too fat.

instrument given over to the three main parts of a bore: the tapered "mouthpiece," the cylindrical center section, and the bell, which starts out smoothly with a slowly increasing taper and ends with a rapid flare.

	TRUMPET	TROMBONE	FRENCH HORN
Mouthpipe taper	21%	9%	11%
Cylindrical part	29%	52%	61%
Bell	50%	39%	28%
Over-all length	58.6 inches	104 inches	208 inches

These fractions are all computed for the "open" instrument; i.e., the slide is all the way in, or the valves are untouched. There is a considerable change when the slide is all the way out or the three valves depressed. For example, the three numbers for the trombone become 6%, 68%, and 26%, respectively.

I have drawn Figure 43 to show you just how the vibration recipe of the trumpet compares with that of a trombone playing the corresponding note in its scale. Notice the resemblance of these recipes to that of our imaginary instrument. The heights of the lines that stand for the amounts of the various ingredients are not drawn here to indicate vibration amplitude, or even sound energy output. Instead, they are drawn on the electrical engineer's decibel scale, which happens to give us a rough indication of the relative loudness of the various components. To our unpracticed eyes, both recipes look almost the same; playing trombone music that was recorded at 33 rpm, on a 78 rpm turntable (so as to raise its pitch by about fifteen semitones into the regular playing range of a trumpet), should give us a good imitation of trumpet music. No musician would be fooled by such a trick,

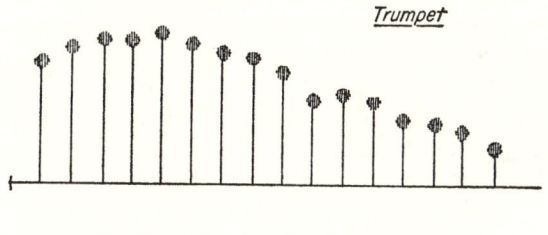

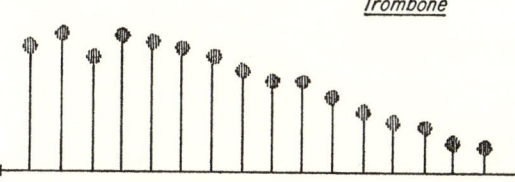

Fig. 43. Sound recipes for the trumpet and trombone. Notice their resemblance to each other and to the lower line in Fig. 40. For both instruments there is appreciable loudness even out of the sixteenth harmonic. (Based on data from Dr. E. L. Kent, of C. G. Conn Ltd.)

nor by the reverse one of playing trumpet music at low speed, but this is partly because of the differences in the way the two instruments begin and end their notes, and chiefly because the slide gives a different approach to each new pitch than does a moving valve.

Having discovered that a trombone and trumpet are similar in tone color, we are stuck with the problem of explaining why a French horn can sound so unbrassy as to be allowed into refined association with woodwinds in a quintet. One big reason is the difference in mouthpieces, but another comes from the fact that French-horn notes are all based on high vibrational modes of the bore. As a result, almost all

193

the ingredients of the sound recipe are high enough in frequency for effective radiation by the bell, and the recipe for the sounds we hear does not differ so much from the recipe within the bore as it does in more ordinary brasses. Figure 44 gives a typical vibration recipe for a gently played French horn, and

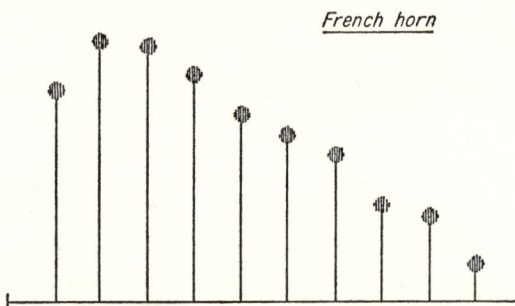

French horn

Fig. 44. Sound recipe for a French horn. While the general shape of the recipe itself is somewhat like that of the trumpet and trombone, the loudness of the higher components falls away much more rapidly. (Based on data from Dr. E. L. Kent, of C. G. Conn Ltd.)

you can see that the lower three or four components behave somewhat as those of the trombone and trumpet do, but the loudness for the higher components falls away much more rapidly.

Let me now summarize the chief things we have learned about the brass family of musical instruments. All are tapering tubes of brass supplied with cup-shaped mouthpieces into which the players fit their lips. The vibrational modes of the brass tubes are carefully arranged so that some of them are whole-number multiples of a certain frequency which is called the pedal note. It is not necessary that there be an actual bore resonance at the pedal note, since it can always be played as a privileged note. It is also possible in principle to have any number of unused vibrational modes of the bore lying above, below, or even mixed in among those forming the harmonic series used for playing. Valves or slides are a means of adding to the length of the bore in order to give new notes filling the scale of harmonics to which the bugle is restricted, but these valves and slides bring in complicated tuning problems whose solution requires great ingenuity and perhaps a well-trained lip. The tone quality of a brass instrument is characterized by a vibration recipe in which the loudness of the first few ingredients increases relative to the lowest frequency component, while the higher-frequency ingredients decrease smoothly in loudness. Designers of brass instruments have a certain amount of freedom in choosing the shapes of their bores to satisfy the twin but partially conflicting requirements of proper tuning and suitable tone color. For this reason the measurements of instruments made by different manufacturers will often differ startlingly from one another, even though they sound about the same when played.

CHAPTER IX

Woodwinds

Our musical wanderings have brought us at last to the regions in which my own first travels began, the land of woodwind instruments, which is populated by sticks of wood with silver keys sprouting from them. The natives of this land have enslaved a dedicated but harried tribe of musicians who spend nearly as much time tinkering with delicate mechanisms or scraping and soaking reeds as they do in playing their instruments.

Reeds and Mouthpieces

"Flutists and brass players, with all their worries, at least have the advantage over players of reed instruments in that their sounds are generated with the help of the solid unchangeable material of the instrument. Oboists, clarinetists, and bassoonists are entirely dependent upon a shortlived vegetable matter of merciless capriciousness, with which however, when it behaves, are wrought perhaps the most tender and expressive sounds in all wind music." This is how

Anthony Baines opens the chapter on reeds in his thoughtful and authoritative book, *Woodwind Instruments and Their History*.

There are two classes of reed systems in use today. One is the so-called single reed familiar to clarinetists, saxophonists, and certain lazy oboists, and the other is the double reed commonly used on oboes, bassoons, and bagpipes. The upper part of Figure 45 shows a clarinet single reed beside its mouthpiece, as well as a cross section cut through an assembled reed and mouthpiece. Blowing with proper lip tension bends the reed slightly toward the nearly flat upper surface of the mouthpiece in such a way as to partially close the aperture into the interior of the bore. The air rushing through this aperture tends to close it still more, and only the peaks of the pressure fluctuations caused by the air vibrating in the bore push the reed open momentarily once in each cycle. The phase relation between the reed's motion and the pressure variations in the bore has already been discussed in Chapter VII, where we found that in order to have sustained oscillations, the frequency of vibration when the reed is blown without a bore *must* be higher than that of the note desired from the assembled instrument.

Every clarinetist knows that a reed attached to a mouthpiece alone gives quite a high-pitched sound, but this is not a proof of the correctness of my statement, since the cavity within the mouthpiece itself constitutes a bore of noticeable length. If you have an old and otherwise useless mouthpiece, you might find it interesting to cut away as much of it as possible, to see how high a note can be got, and thus get an idea of the blown frequency of an isolated reed.

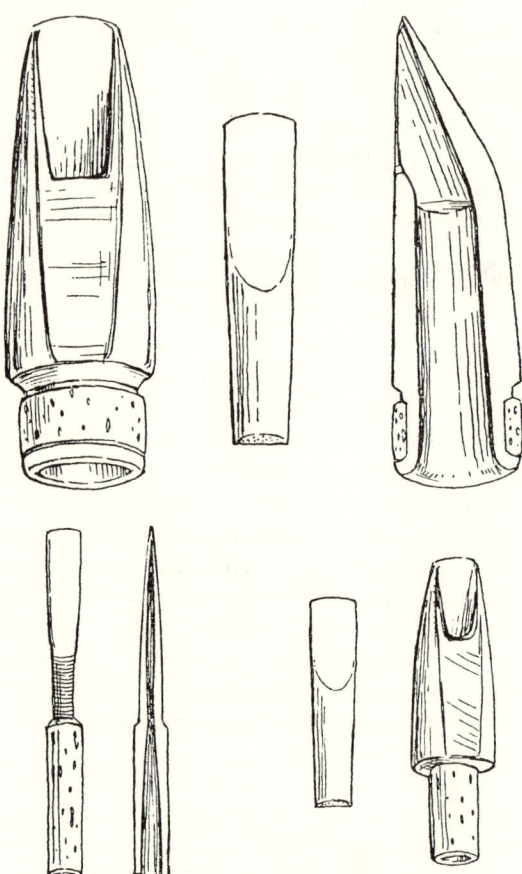

Fig. 45. Single and double reeds for woodwinds. Top: Mouthpiece, reed, and cutaway mouthpiece for clarinet. Bottom: Oboe reed systems, including a clarinet-type single reed.

Much of the effort spent in reed making goes into adjusting the thicknesses of various parts of the reed to make it "roll" open and shut on the slightly curved "lay" of the mouthpiece in a way that generates the correct shape for the puffs of air that enter the bore. It is the vibration recipe of these puffs that ultimately determines the tone of the instrument, just as in the brasses. We find also that the shape of the cavity within the mouthpiece has a great deal to do with the tone; it affects the frequencies of the highest vibrational modes of the bore, of which the cavity is really the uppermost part.

The oboe players' simple-looking double reed appears complete and in cross section, on the left in the lower half of Figure 45. A carefully gouged strip of cane is doubled over and tied tightly to a small and slightly conical brass tube called the "staple." Then the folded end is cut free and scraped thin to provide the vibrating tip. If you look end-on at this tip, it has the appearance of two parentheses close together so they nearly touch (). When this reed tip is pressed between the musician's lips, the edges do touch, and while he is playing, the vibrating reed closes the air passage by flattening out during a part of its cycle. Since both sides of the cavity under the reed are symmetrically made of elastic material, changing lip tension not only alters the curvature and shape of the reeds themselves, and so affects the nature of the air puffs they generate, but also can change the volume and shape of the cavity itself considerably. This mechanical flexibility in the reeds gives the oboist his great musical flexibility in pitch, loudness, and in tone color, and contributes greatly to his miseries if all does not work well. We have at last found a situation in which the physicist's adjective (flexible) for a mechanical system agrees with the same adjective as used by artists to describe something which is important to them!

I shall close my disquisition on reeds by remarking that a *suitably proportioned* reed of either class can be used on any of the woodwinds, without appreciably altering the characteristic tone color of the instrument. That is, an enlarged bassoon double reed can be attached to a clarinet or saxophone, a saxophone mouthpiece goes well on a bassoon, and a miniature clarinet mouthpiece is commercially made for use with oboes. An example of this last appears in the lower right of Figure 45. No self-respecting oboist with serious intentions will be caught dead with such an immoral device, but the jazz saxophonist who must occasionally pick up an oboe on short notice during a performance soon learns that a single reed and properly designed mouthpiece will pull him through safely, and very few of his listeners will notice the difference. This book is not intended as a catalog of musical short cuts, but the possibility of getting *nearly* the same tone from two radically different kinds of reeds is interesting and important to musical physicists.

Woodwinds vs. Brasses

From the point of view of an unprejudiced physicist, the family of woodwinds is distinguished from the brasses in only one or two small-looking but musically important ways. First of all, instead of using their own thick and heavy lips as the reed-valve mechanisms, woodwind players use little splinters of

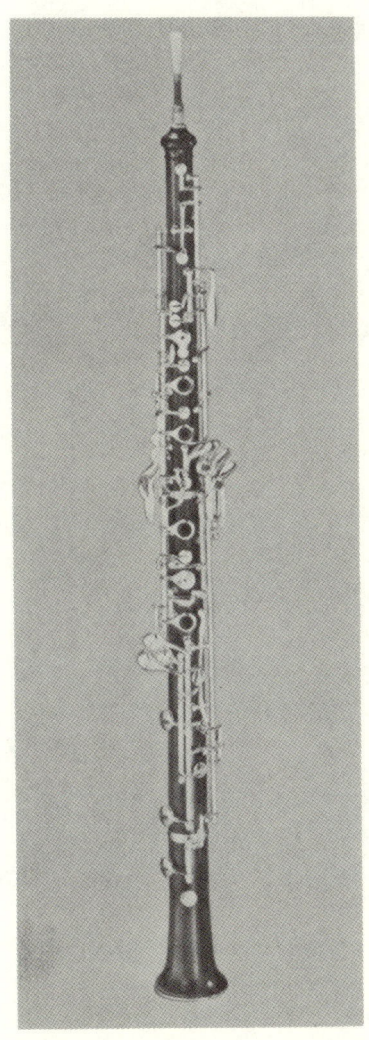

Plate V. Oboe (Courtesy C. G. Conn Ltd.)

Plate VI. Saxophone (Courtesy C. G. Conn Ltd.)

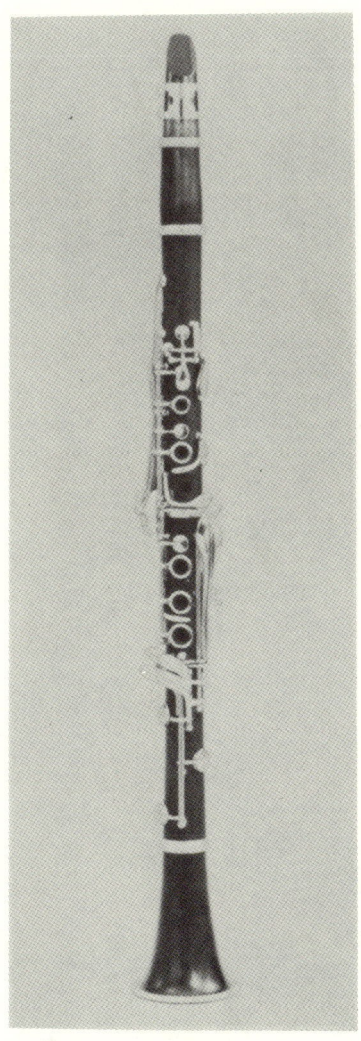

Plate VII. Clarinet (Courtesy C. G. Conn Ltd.)

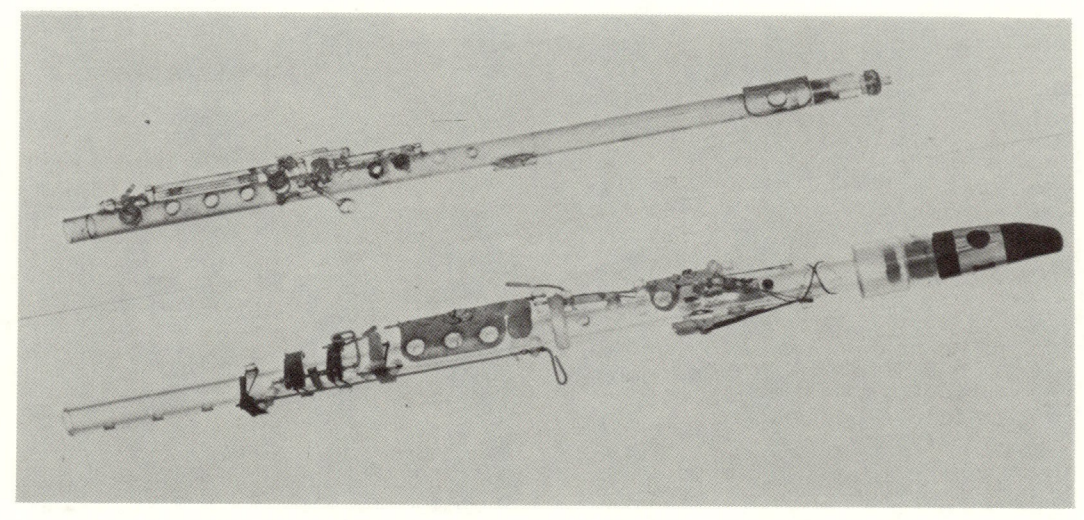

Plate VIII. The clarinet in C shown at the top was built along the lines described in the text (Chapter X). Although the fixed hole size sharpens the tuning of the higher registers a bit, the instrument gives a complete chromatic scale over the whole range of an ordinary clarinet. Gregory Levin built this clarinet, and Stuart Hirsch played it for him as a joint entry in the 1959 Northeastern Ohio Science Fair.

cane carefully shaped and soaked. We found that while the vibrating lips of a trumpeter are strongly *influenced* by what goes on in the bore, they are also able to play a family of privileged notes, these being only secondhand and perfunctory admissions by the lips of their debt to the horn. The family of wood-winds, however, is one in which the bore has an al-most total domination over the action of the reed, at least as far as pitch is concerned. By this I mean that frequencies close to those of the vibrational modes of the bore are the only ones that can be excited, and we find it hard to make a woodwind change back and forth from one of its possible vibrational modes to another by means of lip pressure alone. As a result we can use only the first two or three modes and must help the transition from one to another with certain little tricks with the side holes.

If you have a clarinet handy, try to squawk it in the first half dozen modes of vibration of its bore, keeping all the holes closed. It is possible to eke out a semblance of bugle music from the higher few modes, which are the same ones which plagued the bugler of my fairy tale. You will probably find that your reed has been so insulted by all these goings-on that it will refuse to play refined music any more.

We found that the slide or valve system of a brass instrument needs to shift the natural frequencies of the horn over a range of only seven semitones in or-der to give a complete chromatic scale. The reason for this, of course, is that the musical intervals be-tween the successive vibrational modes of the horn are nowhere more than this distance apart over the musically useful range. Let us see why we could not play the trick again and make a woodwind that plays like a trombone.

I have remarked already on the difficulty of changing from one vibrational mode to another; a trombone with an ordinary clarinet mouthpiece or bassoon reed gives you an exceedingly "stiff" and unobliging instrument whose only virtue is that it will show you that the tone quality is quite brassy, even when played with a reed. Because of this unwieldiness, we find ourselves restricted to the bottom two or three modes, which are the most widely spaced ones with regard to pitch. You will say that all we need to do is to arrange a slide or valve system covering an octave (twelve semitones) on a truly conical bore, or a twelfth (nineteen semitones) for a uniform pipe. This is very simple to say, but you must think of the implications of your remark. First of all, the player of a slide-type instrument would have to learn not just seven but somewhere between eleven and eighteen positions! Now, musicians seem to be able to learn practically anything they have to in the way of finger position and lip tension, so this is not an overwhelming problem. But the thought of playing any music with large skips in it becomes a nightmare; you would kill your musician with calisthenics if he were required to attempt a lively bugle call!

There is a more fundamental difficulty in the physics involved, one that we met in the last chapter. Altering the slide length on a cylindro-conical bore alters the *shape* of the bore cavity; the vibrational frequencies beloved of the bore have different ratios one to another when the slide is in than when it is out. We found that this poses a difficult enough problem in the tuning of regular brass instruments, even with a mere seven semitones' worth of additional tubing to contend with. Our new-type woodwinds would have this problem magnified several times by the wider pitch interval to be covered. The only bores that are strictly free from this particular difficulty are the completely cylindrical ones, which require so much slide, and the true cones, for which no slide can be made. We are led therefore to the realization that woodwinds made in imitation of the plumbing of a brass instrument are doubly impractical, for they cannot usually be built in tune, and it would be beyond the ability of a player to negotiate leaping passages in his music.

Finger Holes

The brass players have driven us away from our attempt to invade their territory with cane reeds, so we must turn to another possibility for getting different lengths of horn. Drilling holes in the side of the pipe will allow us to have our cake and eat it, too, with regard to tuning and agility.

Suppose we exercise our logical powers a little to deduce what is the effect of a hole drilled through the side of an instrument bore near the open end. We use the fact that the frequency of vibration of the lowest mode increases as the bore is shortened, and the fact that in nature humanly perceptible effects usually change smoothly along with their causes from one extreme to another. When the hole is just a few thousandths of an inch in diameter, we should expect the pitch of the note to remain nearly unchanged, and when the hole is huge enough that the end of our instrument is ready to fall off, the pitch would naturally be pretty much the same as that of a bore that extends only to the position of the hole. On the basis of this sort of thinking we can guess the correct general behavior that is observed in experiments, a behavior that is illustrated in Figure 46 for a pipe of length L provided with a side hole at a distance D from the closed end.

The top row in the diagram shows this pipe with different sizes of holes, increasing in diameter from left to right, while the second row gives the lengths of shortened pipe which would sound in unison with the pierced ones directly above. For future convenience, let us call these shortened pipes "equivalent" to their pierced mates. We can then speak of a pierced pipe as having a certain "equivalent length" L_e, which is that of its corresponding simple pipe. Sometimes it is useful to speak of the "hole correction" C, which is the difference between the equivalent length and the distance from closed end to the hole. We can summarize our thoughts on holes by saying that the equivalent length L_e always lies between the pipe length L and the length D from closed end to the hole. A minute hole leaves L and L_e almost alike, while a huge hole makes D and L_e nearly equal. Similarly, the hole correction is very close to L minus D for small holes, and falls to zero for large holes.

For the purposes of making a first visit to side holes as they are used in woodwind instruments, we can, on the basis of our deductions, think of the hole as a means for shortening the pipe. The extent to which the pipe is thus effectively shortened depends on the relation of the area of the hole to the cross-sectional area of the bore at that point, and it depends also on the thickness of the wall through which the hole

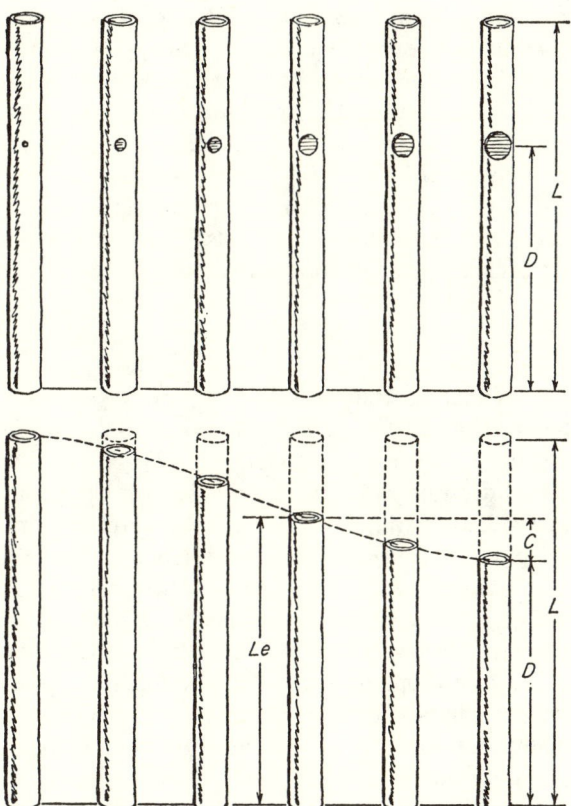

Fig. 46. The size of a side hole affects a pipe's vibrational frequencies. The extent of the effect is illustrated here by comparing different hole sizes with pipe lengths that give matching frequencies. Each pipe in the lower row has a frequency matching that of the holed pipe immediately above it.

is drilled. As a general thing, a pipe with a side hole *does not have the same set of vibrational mode frequencies* as does its "equivalent" shortened pipe. We can think of the process of drilling a side hole as resulting in a shortened horn of a *new shape*. If, however, the hole is drilled close to the pipe's open end, or is followed by a lot of other open holes between it and the open end, then we are safe in assuming that the altered system thinks of itself as having *nearly* the same *shape* as the original undrilled bore.

Getting a Scale

Now that we know a little about the ways of side holes in pipes, we are in a position to see how to use them to give us a musical scale. For simplicity, I shall describe an instrument that is most closely allied with the clarinets, a straight and uniform cylindrical pipe attached at one end to a mouthpiece upon which is clamped a single reed of the proper size. Blowing on the reed will start the air in the pipe vibrating in its lowest mode, and if we drill a succession of eighteen properly placed holes up along the pipe, we can play a rising chromatic scale until we get to a note that is exactly a semitone lower than the pitch of the second vibrational mode of the whole pipe. Our scale can be made to continue if we close up all the holes and shift our manner of blowing to the one that excites the second mode. Once again a successive opening of holes will raise the pitch semitone by semitone in the second mode of pipe vibration. You see from this that we are doing exactly the same thing with woodwinds to fill in the gaps between the vibrational modes of the bore as we did in the brass

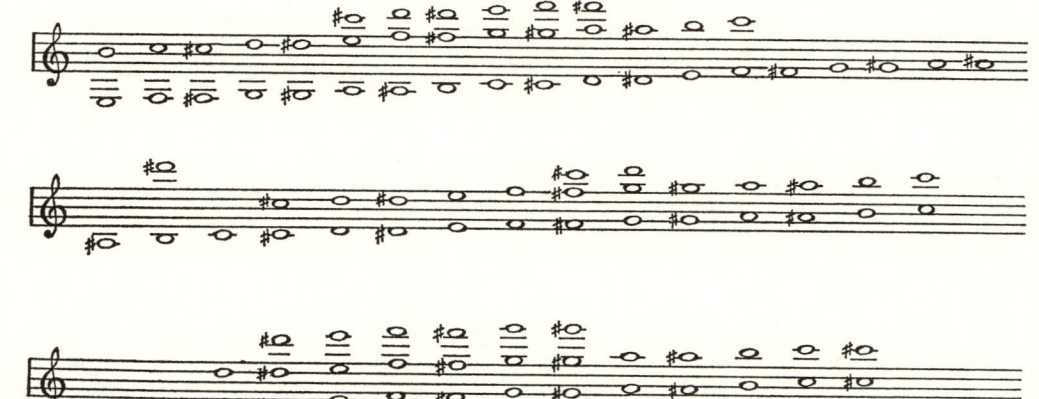

Fig. 47. *Musical scales for clarinet, oboe, and flute. The low register is played using the first vibrational mode (lowest notes on each staff). The scale is continued by using the same fingering on the holes with the reed going in step with the second mode (second line of notes). Many of the highest notes are played by making use of the third mode (top notes on each staff). For practical reasons certain of the notes in the higher modes are not used.*

instruments, except that here we alter the equivalent pipe length by opening holes rather than by adding sections of pipe by means of valves or by a slide. In Figure 47 we have an indication of the way the musical scale is obtained on three of the familiar woodwinds—clarinet, oboe, and flute. On each musical staff, the lowest row of chromatic notes represents the written pitch for each of the successively shortened bores produced by opening side holes one by one, while the instrument is played in its lowest vibrational mode. The upper notes on these same staffs show what is obtained by using the identical number of open holes as before, but with the reed operating on one or another of the higher modes of the air column.

For example, on a clarinet, the lowest mode notes extend from the written E below middle C on up to A♯ above it; musicians call this range of notes the "low" or "chalumeau" register. Coaxing the instrument into its second mode ("middle" or "clarion" register) gives a sequence of notes beginning with B above middle C, which has a frequency three times that of the low E on which we started with the same fingering, and continues steadily up to the C that is written above the staff. For certain practical reasons the remaining higher notes in the middle register are not usable. We get the next note (C♯) by exciting the third mode of the pipe, as is shown in the top line of notes. This third mode ("high" register) vibrates, of course, at a frequency five times that of the first mode, for each effective tube length. I have not followed up the scale beyond high F♯ because the complications of the reed and our simplified version of the action side holes make it impossible to trace out the "parentage" of these higher notes. For a clarinet, however, people usually consider it possible to use notes up six more semitones to C, and the instrument has a really tremendous range of pitch.

Useful Shapes of Bore

I had played various woodwinds for years before realizing that my musically familiar instruments implied an interesting question: what bore shapes can be used in woodwinds? That is, what sorts of bores can be cut off successively at the large open end in order to raise them in pitch without altering their essential proportions at the same time? After a little scuffling around trying to find a proper way to formulate the question mathematically, I was first able to *guess* the nature of a whole family of suitable bores, and later on managed not only to verify the guess but to prove that there is no usable bore other than members of this group, which sometimes are called Bessel horns.[1] When I had found the family of horns with this one necessary property, the problem remained of finding how many of the family satisfy the other requirements on a woodwind bore. I had a fleeting vision of discovering an entirely new kind of musical instrument among the large Bessel tribe, but this quickly faded, leaving the equally interesting realization that there can be no shape other than those

[1] F. W. Bessel was a nineteenth-century German astronomer, whose work on the theory of orbits led him to discover an equation whose mathematical cousin describes the motion of air in a Bessel horn. Here is a beautiful example of the way different branches of science support each other in sometimes quite unexpected ways.

now in use! This comes about very simply: a number of musical reasons (we have met some of them already) require that the frequencies at which the air column likes to vibrate must be integral multiples of the frequency of the lowest mode of vibration. These reasons have partly to do with getting a good tone quality, and partly to do with the proper control of the reed by the bore. (The privileged notes must coincide with the natural vibration frequencies, among other things.) Once we know that the vibrational frequencies must be related by whole numbers, we need only pick out from among the infinity of Bessel horns those few with this second property.

Four members of the Bessel family are diagramed in Figure 48, one of which you will recognize as the familiar cylindrical pipe, with frequencies 1, 3, 5, 7, etc., times that of the lowest mode of vibration, while another turns out to be the simple cone whose frequencies are 1, 2, 3, 4, etc., times its lowest mode frequency. These two familiar ones are the only Bessel horns with integral multiple frequencies, and therefore are the *only musically useful bore shapes for use in reed-driven woodwinds*. There is a similar type of analysis which applies to the reedless orchestral flute, but this instrument stands alone as a type in the orchestra, so I shall say little about it. There is one more horn which at first glance appears useful, but it flares so rapidly (the cross-sectional area increases by a factor of 128 for each doubling of the distance from the small end!) that it could not be built, and anyhow the upsetting effects of a reed system attached to it would utterly destroy its magic properties.

In the orchestra, it is the oboe, the saxophone, and the bassoon that are members of the conical-bore type of musical Bessel horn, as you can see from Figures 49, 50, and 51. In the first two drawings I have exaggerated the diameters so that they are five

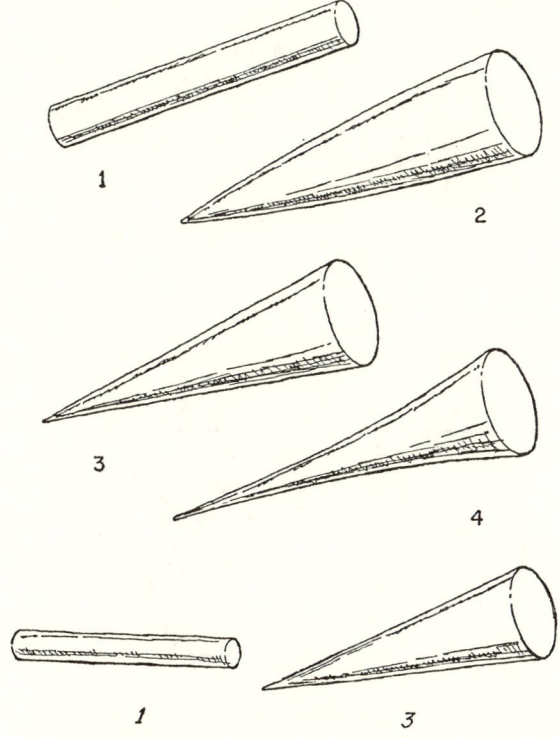

Fig. 48. Some members of the Bessel horn family, which all must be closed at the left end. The first and third of these (shown again below) are the only members of the whole family that are useful in woodwinds.

211

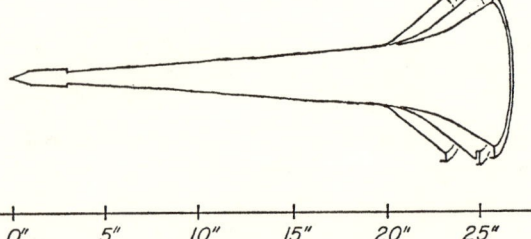

Fig. 49. *Three typical oboe bores. Different manu-facturers use different styles of bells. Diameters are magnified five times. The enlargement at small end consists of staple and reed. (One bore courtesy of C. G. Conn Ltd.)*

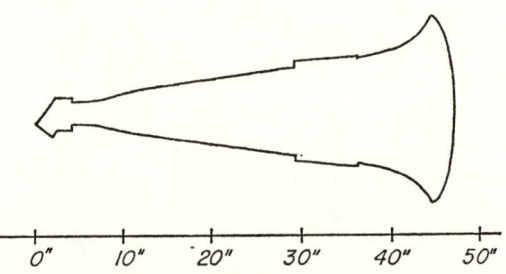

Fig. 50. *Saxophone bore, diameters enlarged five times. Notice the necking-in near the mouthpiece. The mechanical necessities of making the bore turn a sharp corner require the jog near the bell. (Courtesy C. G. Conn Ltd.)*

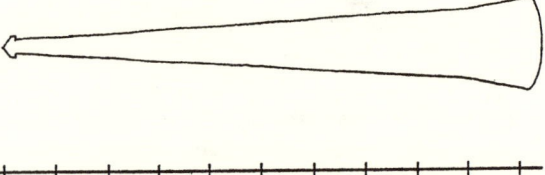

Fig. 51. *Bassoon bore. The diameters are exaggerated by a factor of 10 relative to the length of the instrument.*

times their correct proportion to their lengths; this gives a fat-looking diagram, but does show a little more detail than would otherwise be possible. Because the bassoon is such a long, narrow instrument, it has been fattened up even more—it is ten times too wide. You can see in all three that the bore is really quite a good imitation of a cone, cut off at the small end and provided with a reed. There is more or less of a flaring bell at the open end, but this is in the nature of a correction for certain properties of the holes, and as such I shall explain it later on. In all, the side hole nearest the open end is drilled in the smaller part of the flare, or even above it, as you can see from the photographs in Plates V and VI.

If you are a stickler for mathematical exactness, you will begin to feel slightly swindled at this point, since our instruments are not the complete cones demanded by theory as it has been explained to you. The actual bores are not merely cut off before they reach a point, but there is also a cavity at the small end, made up of the reed and mouthpiece combination. However, a closer look into vibration physics shows us that a conical bore closed by a smallish cavity behaves very much like the ideal complete cone except for one thing. Each of its vibrational modes is lower in frequency than that of the corresponding cone, in an amount that increases rapidly for the higher-frequency vibrations. In the saxophone this effect is compensated to a great extent by "necking in" the bore in the region next to the mouthpiece;[2] in oboes and bassoons, however, the problem is

[2] This "necking in" also serves another, much more obvious, function. It permits the mouthpiece to slip on over the top of the metal cone, leaving room for the cork sealing joint.

solved in a different, and almost accidental, way. Here the player's lips press the two flexible reed halves together so that the cavity shown in my diagram is actually reduced considerably. Pinching gently on an oboe reed with your fingers next to the player's lips while he sounds a note will show you how much the pitch is raised when you flatten the reed system, or lowered when you pinch it edgewise to round it out again. (Your oboist will be wise to allow such experiments only on a reed which no longer commands his esteem!)

The clarinet family, of which an example is shown in Plate VII, has an essentially cylindrical type of bore. Three common clarinet bore shapes are diagramed in Figure 52. Once again I must ask you to defer your questions about the bell, which is quite prominent in one version of the instrument, but we can notice that the mouthpiece cavity forms very nearly an extension of the main bore, and we have little to worry about on its account. I merely say in passing that the irregular shape inside the mouthpiece

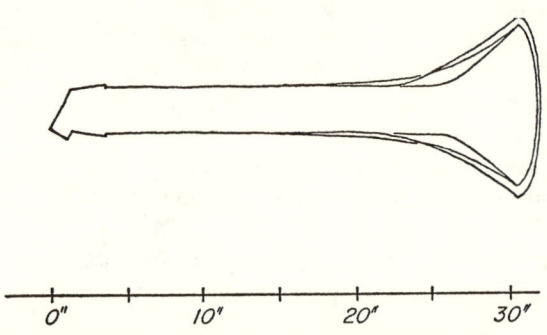

0"   10"   20"   30"

Fig. 52. Typical clarinet bores. Again, the instrument's diameters are shown five times too large.

214

has a noticeable effect on the tone quality of the instrument, and it does so by altering the frequencies of the very highest of the vibrational modes belonging to the bore. To be sure, an enlarged mouthpiece cavity can lower the pitch of the vibrational modes, but they are lowered equally, except for the highest ones. We see that cylinder-bore woodwinds are not nearly as sensitive to the size and shape of the mouthpiece cavity as are their conical-bore cousins in the orchestra.

More about the Side Holes

We have paid a superficial visit to the side holes on a woodwind, and it was sufficient to show that they can be used as a means of effectively shortening the bore. Now that we have discovered certain requirements that the bore itself must meet if it is to be useful in the orchestra, it is time to go back and

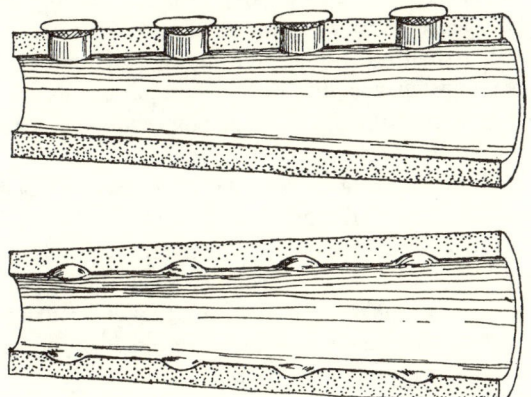

Fig. 53. Closed side holes make the bore "lumpy."

215

see what restrictions have fallen as a result upon the holes.

Let us start out by noticing that even the "unused" closed holes in the upper part of the bore have an effect on the bore shape. We see in Figure 53 a section of woodwind having closed holes, along with the equivalent "lumpy" cavity, whose set of vibrational frequencies obviously would be different from those of a smooth bore. The mathematics of the problem is far beyond us here, but we can learn the answer: if the diameters of the closed holes are properly related to their spacing, it is possible to make the lumpy bore have the same set of vibrational frequencies as that of a smooth bore. The side holes on a woodwind get smaller and smaller as we go up toward the reed, according to a certain mathematical law, in order to keep the bore and its holes behaving as a proper cone or cylinder should. The stripped saxophone of Figure 54 shows very clearly the variation of hole size I am speaking of, while the similarly stripped flute appearing in Figure 55 looks as though I have been contradicted! I shall not explain the flute to you here, but once again tell you that the bore requirements on a tube that is essentially open at both ends (as in the flute) are different from, and much less restrictive than, those applying to the other, reed-excited woodwinds. This is one reason why Theobald Boehm, who made such a success of his attempt to use uniform-sized large holes on a flute, failed to persuade oboe and clarinet makers to accept his thoughts on hole sizes, even though they did take over his ideas on key machinery with great eagerness.

We can now look at the placement of the holes to make them "cut off" the bore at the proper lengths

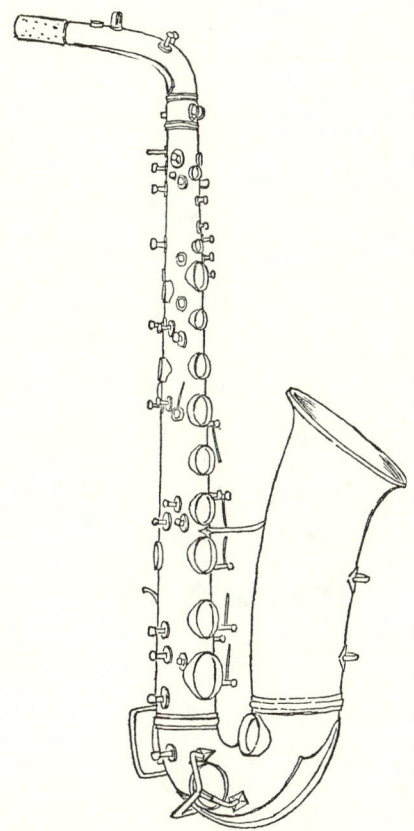

Fig. 54. In a stripped saxophone the variation in hole sizes is apparent. (See also Plate VI.)

and give us a chromatic scale. By the late 1920s physicists had learned to adapt many newly developed methods of electrical engineering to the calculation of vibration problems with such great success that today there are people who have almost forgotten there is

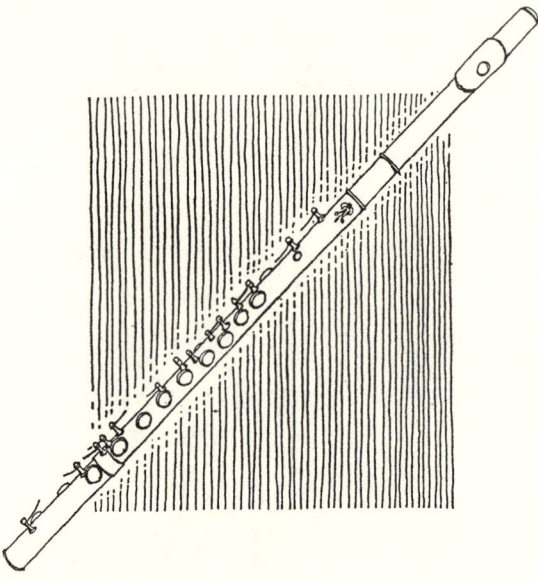

Fig. 55. The holes in a flute are seen in this stripped drawing to be of essentially the same size. Flutes are not subject to quite the same bore and hole requirements as the other woodwinds. (See also Plate VII.)

any such thing as straight vibration theory. They have to translate everything into its electrical analogue, and a great deal is often lost in the process. Be that as it may, properly applied, with imagination, electrical methods are often of great use. Without going into detail, I can tell you that vibrations in the air contained in a tapered bore behave very similarly to electrical oscillations in a telephone cable whose wire spacing varies uniformly. A closed hole in the side of a pipe translates to an electrical capacitor connected between two wires of the cable, while an open hole becomes an inductor (coil) connected across the wires. Using this for a dictionary, you can see that I can translate a typical woodwind bore into its corresponding "tapered line" shunted at the hole positions by capacitors or by inductors. About 1929, E. G. Richardson showed by such methods how to calculate the position of a single hole to give the desired "equivalent length" that would play a certain single note, and while his method can in principle be extended to a system with several open holes, each affecting the rest, in practice the process is too cumbersome to be usable. Manufacturers of instruments were already in possession of a large body of rule-of-thumb knowledge which permits them to make excellent instruments, so that Richardson's work has lain almost unexploited.[3]

The difficulty with Richardson's approach is that each hole is first treated separately from the others, after which their mutual effects must be reconciled. I should like now to describe a means for escaping, or rather bypassing, the difficulty. Thirty years ago telephone engineers, led by W. P. Mason at Bell Laboratories, developed an extensive theory for telephone lines across which many electrical components (capacitors and inductors) were connected *at regular in-*

[3] Richardson describes his calculations in an appendix to his excellent little book, *Acoustics of Orchestral Instruments,* London: Arnold, 1929.

tervals. We very quickly realize by looking at any woodwind that the side holes are *not* spaced along the bore at equal intervals. As a matter of fact, the spacing decreases roughly 6 per cent for every semitone you go up the scale; telephone-line theory is not directly applicable. About four years ago it occurred to me that woodwind players are familiar with something containing the clue to the problem, a clue that reduces the finger-hole problem almost to the status of offhand knowledge of telephone engineers.

Suppose I play a note on some instrument that requires the lowest five finger holes to be open, in the manner sketched in the top part of Figure 56. If now I close the second highest open hole, the note will go flat in pitch, and have a stuffy sound to it, because

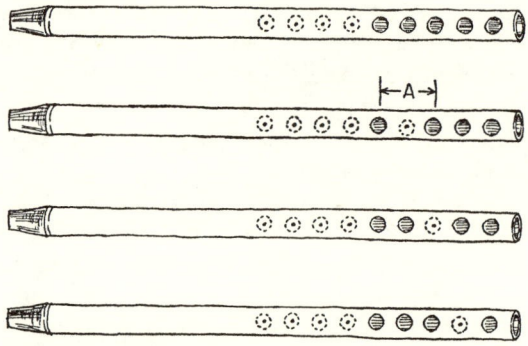

Fig. 56. The effect of closed holes on tuning. If the instrument is designed to be in tune when all the lower holes are open (top diagram), then closing one (second diagram) will flatten the note by alteration of interhole spacing A. Third diagram from top shows fingering that produces only slight flattening of pitch. Flattening is negligible in bottom diagram.

the spacing between the first two open holes has been increased, as you can see from the second line of the figure. If instead I close one or another of the lower holes, one at a time, as shown in the lower lines of Figure 56, we find that the upsetting effect of closing a hole decreases the farther away the closed hole is from the topmost open one. From the point of view of physics, we draw the following conclusion from this and from related experiments: the size and spacing of holes that are more than two or three away from the highest open hole do not appreciably affect

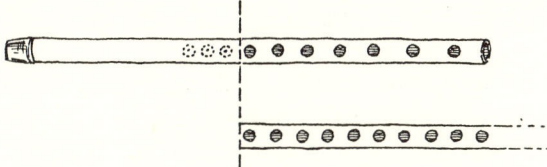

Fig. 57. *Open but variably spaced holes of a real instrument* (top) *could be replaced with a set of holes* (bottom) *uniformly spaced to match the first two of the real instrument.*

the pitch of the played note. (In the high register of an instrument things don't work quite so simply, as we will learn shortly.) Why is this worth noting? Simply this: in my mind's eye, I am free to do anything I please to the lower holes! In particular, I want to imagine them all of a uniform size and spacing, which has been chosen to agree exactly with that of the *first two open holes* of my real instrument, as shown in Figure 57. Such a replacement will not alter the pitch of the particular note, but it does allow me to borrow all the formulas worked out by the telephone engineers that apply to uniformly spaced capacitors and inductors. The first test of my little inspiration was to make a calculation of the frequencies of notes played on my own clarinet using measurements of the positions and sizes of the finger holes. To my great joy, and some little amazement at the simplicity of it, even my first rough calculation gave pitches that were within a small fraction of a semitone of the proper notes of the scale.

Now, it is one thing to calculate the pitches given by an already existing clarinet, and quite another to predict beforehand just where to drill the holes in the first place. The reason for this is that the calculation is based on the interhole spacing, which must itself depend on the hole position. It turns out, to make a long but not very difficult story short, that even a poor guess at a plausible spacing for the holes will give a reasonably good set of hole positions. If you are a perfectionist, you can then use these positions to find a new set of spacings upon which to base an improved calculation, at which time you are already at the limits of common sense in musical accuracy.

Changing Registers

At the beginning of this chapter we learned that we get the complete scale of the woodwind by using the first, second, and some one of the higher modes of vibration to extend the range provided by the finger holes. At that time, I also made quite a point of the difficulty we find in causing the reed system to change its excitation from one vibrational mode of the bore to another. The time has now come when we can re-visit this part of woodwind territory, with a little more leisure, to see just what a musician has to do when he wants to change registers.

Up till now I have continually passed over the flute as being almost a freak among the other woodwinds, but it has a certain simplicity that will make my present explanations easier than they would be if they were based on clarinets or oboes. A flute is very nearly like a uniform cylindrical pipe, open at both ends, and blown like a pop bottle at one of them. Such a system has nearly the same vibrational frequencies as does a perfect cone of the same length, so that its higher modes are located at 2, 3, 4, etc., times the frequency of the lowest mode; in other words, the pipe will vibrate at any one of the harmonics of its lowest mode frequency. Now a flute *can* be shifted rather easily from mode to mode by altering the velocity and angle of the blown air, but even so it is customary for players to use the finger holes to help them make the change to the highest register in a manner which is directly analogous to that which must be used in the other woodwinds.

Suppose a flutist wants to play the high F above the musical staff. Two possibilities might come to his mind. He can either open the holes to give him the low F and then blow in such a way as to set up the *fourth* mode (and therefore also the fourth harmonic) of this arrangement, or he can similarly finger B♭ and play its *third* harmonic. What he actually does is something quite closely similar to both at once, as you can see from Figure 58. He arranges his fingers as though he were going to use his first possibility, and then opens what would have been the highest open hole of the second method. The results of such a procedure are applicable to most of the notes in the upper register of the flute, and may be summarized as follows: finger the note of the same name in the low

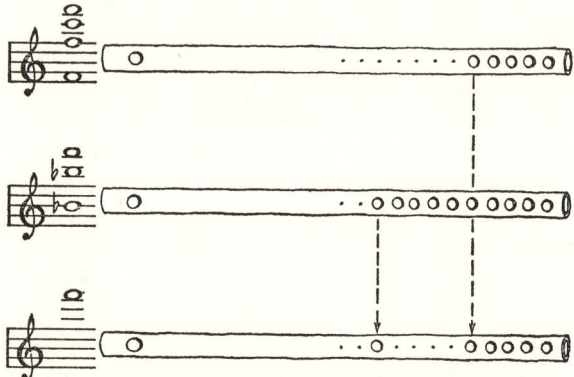

Fig. 58. Top: Fingering for playing the higher F *as the fourth harmonic of the low* F. *Middle: Fingering which gives the same high* F *as the third harmonic of* B♭. *Bottom: Fingering used in practice; notice its relation to the other two.*

register, and open the fifth hole above the highest open hole of this fingering. There are similar but more complex summaries for the other woodwinds. We have here a vibrating system more complex-appearing than we are used to. Theory tells us that the lowest mode of such a system lies at a frequency between those of the two "ancestor" fingerings (F and B♭ in my example), being closer to the upper one if the hole is large, and nearer the lower one if the hole is small. The second and third modes have a similar behavior, with vibrational frequencies that are *not whole-number multiples* of the lowest mode frequency. The fourth mode of the system is just exactly what we were after, while the fifth and most of the higher modes are once again all out of tune. On a large-holed, modern flute, the out-of-tune modes are not very badly out, and can be excited quite easily, but the analogous notes on a clarinet or oboe, or even an old-style pre-Boehm flute, *cannot usually be blown* (try it sometime). The reason is subtle and complex, but once again it is related to the coincidence of privileged frequencies with vibrational mode frequencies, as well as to frictional losses of the air moving in the upper hole. Whatever the explanation, we find that opening holes well up in the bore gives us a way to shift from one mode to another. We can also see very well why a leaky key can cause the squeaks which are so terrifying to wind players.

On a flute, it is so easy to go back and forth between the first and second vibrational modes of the pipe that no provision is made for the change-over in normal playing. An instrument like the oboe, on the other hand, pretty much insists on playing in its lowest mode if left undisturbed, and people have provided three small "speaker holes," which are used one at a time to cause the reed to shift to the second mode. If we take my previous explanations literally, there ought to be a speaker hole for each note of the scale, but in practice this would be hopelessly complex. Each of the three holes on a real oboe is assigned the job of acting with one or another of a group of note holes, and it is placed at some sort of a compromise position. Until one weekend when time permitted a little calculation, I used to worry puritanically about the effect this might have on the tuning of the middle and upper registers. It turns out that if it is small enough, a misplaced speaker hole does not pull the tuning of the desired note very much out of line, even though it can move the undesired lower-mode frequencies far enough that they will not be excited by the reed. As so often happens, the old established custom of instrument makers is justified and correct even when it seems to quarrel with the laws of physics. The quarreling usually arises out of an oversimplified or incorrect application of these laws—or, better, these principles of physics.

I have just said that on an oboe the use of three small speaker holes allows one to get an acceptably tuned middle register, and I have made the point that the acceptability of the tuning depends on the distance of the speaker hole from its "correct" position. How is it then that a clarinet, which has nearly 50 per cent more notes in the interval between the lowest two playing registers, and therefore needs 50 per cent more holes spread out along the length of the bore, can get along with only one speaker hole?

We can reduce our problem considerably by noticing in Figure 58 that only part of the complete first-mode scale is reused in the middle register; the speaker hole has to deal with only fourteen of the nineteen notes, only two more than the twelve notes we have to worry about on the oboe. The rest of the explanation awaits someone's research, although I should guess it is connected rather closely with the wide spacing of the first two vibrational mode frequencies of a cylindrical bore. It is not possible to attribute the difference to the use of single or double reeds, since the single-reed saxophone requires two speaker holes, and a single-reed mouthpiece on an oboe does not reduce its predilection for many little holes.

What Is the Bell for?

In brass instruments, all the air in the bore vibrates while that part of it in the bell communicates these vibrations to the outside air, and thence to our ears. In a woodwind instrument, on the other hand, the presence of side holes makes the situation quite different. We must make a distinction between the way in which sound is radiated when several holes are open and the way it comes out of the bell when all the holes are closed.

Imagine that we have a very long pipe with side holes drilled into it over most of its length (a woodwind with almost all its holes open will do as an example). With the help of our telephone cable equations we find that sound is radiated only from the first few open holes on the pipe (for holes like those on a clarinet, the first two or three holes are the only ones which contribute much to the total loudness), and this radiation is not very "efficient" because of the smallness of the holes. This statement is true for the first half dozen harmonics of an ordinarily played frequency, but all ingredients whose frequency is above a certain limiting value "leak out," so to speak, and are radiated very effectively from *all* the lower holes, and as a result are emitted very efficiently. The exact way in which the change-over occurs between the two sorts of sound emission depends critically upon the size and spacing of the open holes, and this provides us with a definite physical basis for understanding the effect of the hole sizes on tone quality.

You will ask why I seem to have gone off on a tangent dealing with the radiation of sound from side holes, when the heading of this section promises an explanation of the bell. The reason is this. Not all the notes on a woodwind are played with several open holes along the lower part of the bore, and in particular the lowest note of all is played with no open hole at all in a normal instrument. The bell is put on the lower end of the bore to take the place of the missing holes; this it can do by virtue of the radiation properties of bells that we discussed in the chapter on brass instruments. A properly shaped bell can have very nearly the same sort of transition from inefficient to efficient radiation as do the holes it is designed to replace, and thus it prevents the emission of unmusical snarling noises that would come out of the abruptly ended bore.

Once again the flute is different from other woodwinds; it needs no bell for two reasons: first because the vibration recipe of any flute, old or new in style, is so lacking in the higher components that very little sound energy is present above the change-over frequency, and, second, the very large holes of the modern Boehm flute serve to cut off the tube so effectively that *all* the notes think of themselves as coming out of the end of an ordinary pipe!

One of the most interesting experiences I have had in recent years was to hear the change in the tone quality produced by a clarinet mouthpiece that was attached alternately to a pipe having a row of *closed* holes all the way down to its end and a similar though longer pipe with enough holes open at the lower end to make it sound the same pitch. The pipe with several open holes sounded very much like a normal clarinet, while the other one was reminiscent of the dull noises you get when trying to play a piece of conduit as a bugle. It is the side holes which turn a wooden pipe into a musical instrument, and the bell is a portable substitute for an elongated bore with unused but open side holes.

The bulbous or pear-shaped extension at the foot of an English horn does not fit in with my explanation. This instrument, which is basically an enlarged oboe, has the reputation of having a particularly nasal sound associated with the pastorale. The hollow bell forms a sort of cavity resonator, which alters the tone in a manner very similar to the way in which your speaking voice is altered when you talk through closely cupped hands, or with your mouth near a wide-mouthed jar. Careful experiments have shown that the note played with all the holes closed, and to a lesser extent the next two or three above it, is altered by the odd-shaped bell, while the rest of the scale is very much like that of an oboe that has been lowered in pitch. Composers who want to exploit the outstanding distinction of the English horn are obviously limited to a very small range of notes! If on the other hand they want to use the instrument as a solid and useful middle voice in a woodwind choir, they do not find it overly difficult to avoid the peculiar notes; there are alternate fingerings that avoid all but two or three.

Vibration Recipes and Woodwind Tone Color

It is an almost hopeless task to say anything useful and at the same time brief about the characteristic sounds of the various instruments. Physics cannot as yet explain much in detail about the recipes, but we do know why the vibration recipes of conical-bore instruments contain all ingredients that are whole-number multiples of the played frequency, while the cylindrical-bore clarinet in its most characteristic register (the low one) emits sounds in which only the odd multiples of the fundamental are noticeable. This at first sounds like an obvious repetition of my earlier statements about the frequency ratios between vibrational modes of the two classes of woodwind, but the effect is in fact only a descendant of these ratios, and can be shown by an additional argument, which I shall not work out here. Figure 59 gives a set of fairly typical vibration recipes for the oboe, saxophone, and clarinet played in their low register. While measurements of such recipes will differ quite widely among themselves for different players and instruments, we can at least say that the oboe's low register recipe is characterized by harmonics which are stronger than the fundamental component, so that the over-all shape of the recipe is somewhat reminiscent of a trumpet recipe. I recall being thoroughly fooled on hearing a distant trumpet play some music that is usually associated with the oboe. The deception was finally given away by the un-oboish way in which the notes began and ended. The recipe for a saxophone is much harder to describe, except to call it irregular, with several components rather weakly represented.

230

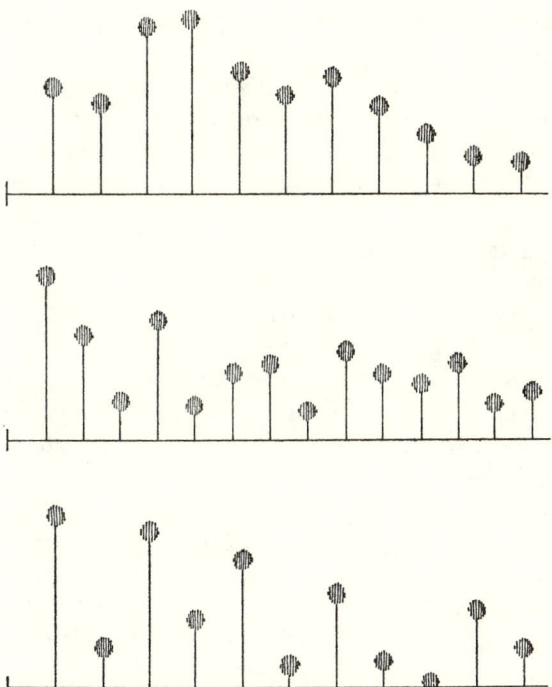

Fig. 59. *Vibration recipes of an oboe, saxophone, and clarinet, all played in the low register. Notice that some of the higher harmonics of the oboe are louder than the fundamental. The even harmonics of the clarinet are considerably weaker than the odd ones.*

In agreement with theoretical arguments, the clarinet recipe has only very weak contributions from the even harmonics of the played note, and it is an experimental fact that, the weaker these are, the more characteristically clarinetish the sound appears to musicians.

We should now turn to Figure 60, in which we find recipes for the middle register of a saxophone and of a clarinet. The saxophone recipe is not drastically different here, although there is a less marked variation in loudness between the various ingredients than there was before. The clarinet, on the other hand, has lost much of its distinguishing characteristic; here we find both even and odd harmonics appearing in the recipe, although the odd ones still possess a little of their prominence. If we had recipes for notes played in the third register of the oboe or clarinet (the saxophone does not normally play above the limits of its first two vibrational modes), we should find that they look very much alike, and indeed they sound pretty much alike as well, barring the subtleties upon which music so often rests.

In conclusion we must always remember that different players will scrape their reeds and adjust their lip pressures to get whatever type of sound recipe they are after, within the rather broad limits set by the fixed shapes of their instruments. The influence of the musician on the way in which his reed opens and shuts the passage between his lungs and the interior of his instrument is so great that a skilled player can get nearly the same sounds (as judged by a listener) on an old piece of junk as he can from his own treasured instrument. He, like violinist friends, may

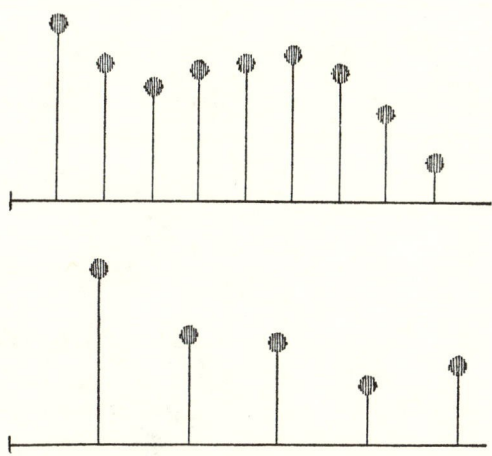

Fig. 60. Saxophone and clarinet recipes for the middle register. Notice that the clarinet has lost most of the characteristic emphasis on the odd harmonics.

be able to do this, but his life is much easier if the bore he uses, and its side holes and mouthpiece, are all of a sort to give him what he wants with a minimum of struggle. He is then free to use his skill in the performance of music rather than in exhausting battle with the vagaries of an unsatisfactory machine.

CHAPTER X

Homemade Wind Instruments

Many people grow up with the impression that musical instruments were created full-grown and sent down to us from heaven in a manner reminiscent of the birth of the Greek goddess Athene. Having read this far, you probably have begun to realize that the only similarity between the development of an instrument and the arrival of Athene is that both begin with the headache of their creator! This final chapter is devoted to brief notes that should help you to make one or another simple wind instrument. I assume that you already can play the conventional instruments that are most similar to these, for it is a pretty hopeless job to coax life into a newborn toy if you have never enjoyed the mature companionship of a properly built and adjusted commercial instrument. I give you only brief descriptions of the work, trusting your ingenuity to fill in the gaps. The suggestions I make are not so much intended to exclude other ways of going about the job as to start you on a line of thought, and to call your attention to the possibilities of wire, tin, solder, and glue.

A Sort of Trumpet

Suppose we try to design a "trumpet" on which you can play most of the notes of the scale without, however, getting into the elaboration of three valves. The whole trick is to have the tube long enough so that only the higher-numbered vibrational modes are used, since these are spaced out at fairly small musical intervals. Our problem is almost precisely the one that faced musicians in the days of Bach and Handel, and we can borrow inspiration from them. The German musician and scholar Werner Menke looked closely into the history of trumpets and wrote a book entitled *History of the Trumpet of Bach and Handel* (London: Keever, 1934). Not only does he critically examine early instruments and their manner of use, but he also describes a modern adaptation of the old design. As some of you know, much of the early trumpet music is next to impossible to play on our customary modern instruments. Our homemade instrument will be a cousin to Menke's design.

Let us start out by assuming we have a piece of ½-inch inner diameter copper water tubing 78 inches long, into the end of which we jam an ordinary trumpet mouthpiece. The mouthpiece will slip right in up to the cup, but I have taken account of this. Just make sure that the mouthpiece taper is wrapped in enough plastic tape to keep the joint strictly airtight. If you fool around with this combination for a while, you discover that a pretty fair scale is available to you if you ignore the notes below piano middle C. The scale is not complete, and we ought to look into the possibility of putting in a single homemade valve to help fill the gaps by giving an additional one-tone depression of the pitch. If we can rig some machinery to do this for us, our available notes at the low end of the scale become the following:

Playing on such an instrument requires you, even more than usual, to "hear" each note in your mind before playing it, but the instrument will become manageable with practice.

Figure 61 shows a simple-minded adaptation of the ordinary rotary valve which can be built for our instrument. When the rotor bar is horizontal, we have the analogue to trombone position 1, and turning it 90 degrees brings in an extra loop of tubing. The general principle of valve construction should be quite obvious from the diagram. The rotor is made of a cutaway piece of tubing (A) in which a rectangular block of wood (B) is held by means of plaster of Paris (C). The nails (D) are pushed through drilled holes in the block after the rotor is slipped into the valve body (E), allowing the rotor to turn without letting it slip out of place longitudinally. While playing, you turn the valve back and forth with your right hand by the projecting end of (B).

The valve body (E) can be made in many different ways; all that is required is a smooth hole for the rotor. A pair of the telescoping brass tubes used on kitchen sink drains might provide the raw materials for (A) and (E). Another possibility for (E) is to

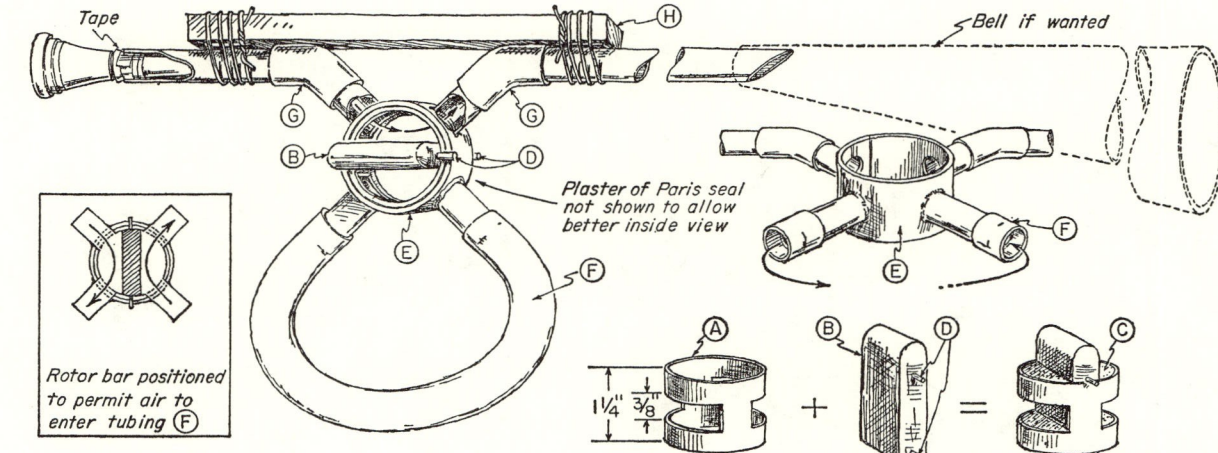

Tape

H

Bell if wanted

G

G

B

D

Plaster of Paris seal
not shown to allow
better inside view

E

F

F

E

Rotor bar positioned
to permit air to
enter tubing F

A

$1\frac{1}{4}''$ $\frac{3}{8}''$

B D + = C

Fig. 61. Diagram of the general method of making a "trumpet."

drill a hole of the proper size in a block of plastic or hardwood, in which case the four stubs of copper tubing must be jammed in and glued instead of being soldered or brazed in place. In order to make life as simple as possible, and to give yourself room for experiment, the additional tubing (F) should be made of a piece of soft rubber tubing, cut so that the valve adds exactly 9 inches to the over-all bore length. Your best bet is to leave it a little long and then trim it down until the musical interval is proper.

Because the bore is very long and the joints at the valve are weak, it would be wise to make use of a pair of 45-degree elbows (G) on the main pipe and mouthpiece to get them in a straight line as shown. Lining it up in this way allows you to clamp a wooden splint (H) across the two sections to hold everything firm. For strength and playing convenience it is a good idea to keep the distance between the mouthpiece and the valve down to about 6 to 9 inches, while your tinkering will be much simplified if the tubing of the main bore beyond the valve is cut off a couple of inches beyond the splint, and rejoined with a soldered coupling sleeve. This allows you to do all the complicated work on the valve without having to worry whether the long tubing will flop around and break things. When the valve is solidly assembled and the splint is in place, it is easy enough to coil the rest of the tubing into a convenient shape (preferably in a way that permits it to be tied to the splint) before it is slipped into the coupling and soldered tight.

So far I have described a bell-less instrument as being by far the easiest to build and tune. If you should want to make a bell, Figure 62 gives a pattern for cutting the sheet metal. A good material to use is the soft copper or zinc-alloy sheeting sold for gutter repairs and chimney flashing. You will find it quite difficult at best to roll the cutout sheet into a neat cone, but a little of your trouble will be saved if you drill holes as shown along the edge of the sheet. Sheet-metal screws can be put through these to hold the

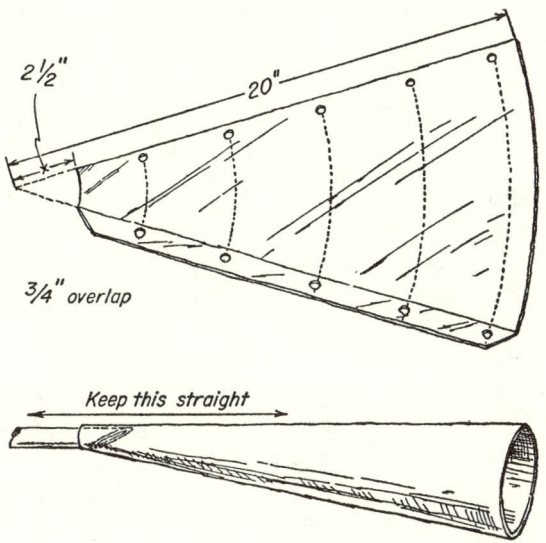

Fig. 62. *Top: Pattern for cutting out a conical bell for the trumpet. Bottom: Suggestion for attaching the bell to the main bore so that it is fairly secure, although crooked-looking.*

edges together for soldering, after which the screws can be removed to make room for the broomstick, bottle, or whatever else you may feel like using as a wedge to help you get the bell looking more or less like a cone. Needless to say, the screw holes must be filled with solder or taped over if the bell is to act in a normal way.

One of the hardest jobs for an amateur tinsmith is the joining of a cone to a pipe. Here, a leakproof strong joint is more important than appearance, and you should use the slightly cockeyed arrangement shown in the lower part of Figure 62. Saw a long bevel off the end of your copper tubing and slip the bell over it. Allow it to hang in such a way that the top of the bell is parallel to the tubing, as shown, and run in a goodly supply of solder. It is a good idea to replace a couple of your sheet-metal screws during this operation, to stave off disaster! Any leaks can be filled in later with Duco cement or covered over with a layer of plastic electrician's tape.

We come now to the trick of retuning. Depending on the exact shape of your bell, the pitch of the trumpet will be lowered by somewhat more than a tone, and you will need to shorten the tubing by 9 inches or so to bring it back to pitch. Since the trumpet is not built in a standard way, however, and almost certainly will not be played in an ensemble, the simplest thing is to leave it unchanged, with the possible addition of a little more hose in the valve loop to compensate for the added bore length.

Woodwinds Simplified

All the woodwinds have many more holes to be controlled than any normal person has fingers, and over the years a number of more or less standard systems of key mechanism have been developed to help with the job. Before we worry about just where all the holes are to be drilled, I must give you a few suggestions for the rudimentary mechanical engineering of machinery. Once again I rely on your own knowledge of commerical practice as a background to my comments, in order to save endless time and detailed drawing. Almost all the keys on a woodwind can be made on the pattern shown in Figure 63, as you can see on the clarinet of Figure 67. Copper or zinc-alloy sheeting with a thickness of about .016 inches is ideal for the folded metal parts; brass or galvanized-iron wire about $\frac{1}{16}$ inch in diameter is stiff enough for the axles and lever arms without being too rigid for easy construction. Aluminum sheet and wire are not stiff enough for this use, and cannot be soldered.

There are two kinds of useful key posts, those shown as (A) and (B) in Figure 63. Style (A), being made of a bit of $\frac{1}{4}$-inch Lucite rod, is easily cemented to a Lucite tube with Duco glue. The more compact post made of folded metal may come off if you use Duco alone, but a thoroughly dried preliminary light coat of Goodyear Pliobond cement will provide a reasonably good grip for the Duco. Be careful not to use too much of the Pliobond or the post will feel rubbery when the joint is finished, preventing smooth action of the machinery. The axle (C) should always be soldered to the key itself to prevent it from wobbling. To hold the wobbling to a minimum, you always want to keep the key posts as far apart as possible (up to a couple of inches) so as to provide a good long base line for your loosely fitted pivots. When needed, a small washer or loop of wire can be soldered onto the axle to prevent end play. It is usually better practice, however, to arrange things so that other parts of the mechanism are placed so as to do

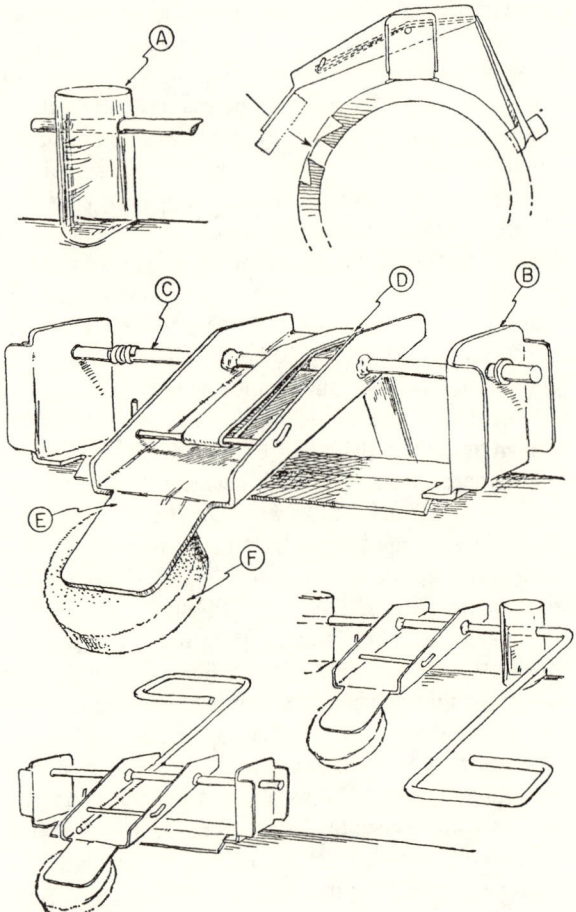

Fig. 63. Basic key design for homemade wood-winds.

away with the need for soldered-on washers, even though loose ones may be used to cut down friction. Rubber bands make very practical springs. To hold a key open, they can be attached as shown at (D) by means of a small bit of thin wire or a nail run through holes in the key during final assembly. The band is then run over the axle and projecting heel of the key, and then on to its anchorage around a Lucite peg glued to the body of the instrument. If the resulting tension is too great for comfort, the leverage may be reduced by cutting away or bending down the heel of the key.

Keys that normally are to be kept closed can be made with the rubber-band anchorage on the heel end, or the rubber can be simply looped over the key and pulled taut in any way that gives the proper action. Many of the keys have to have their tails extended to some location that is comfortable for your fingers, a matter that is easily taken care of with suitable bent-wire loops soldered to the key in the manner shown in the little sketches of Figure 63. Because the materials are light and a little hard to manage, you will find it helpful to put a small drop of acid flux on each spot where soldering is to be done. A good hot iron and some electrician's solder will then give you a quick joint without the need for scrubbing around on the joint to get the solder to wet it. Whenever you use acid flux, it is imperative that the work be rinsed off in tap water as soon as it is finished, otherwise corrosion will disfigure it and may even cause it to jam.

The seating of pads over the holes leads many amateur repairmen to their doom. You must not depend on strong fingers or spring pressure to push an irregularly mounted pad into airtight contact with its seat. Bend the tab (E) around carefully so that a pad (F) glued onto it can rest so evenly upon its hole that only the lightest pressure is required to seal it. It is only a matter of common sense that the pad be glued in place after the key is securely assembled on the instrument, so that it can be laid upon the hole with a small blob of Duco on its back. In this way a light pressure on the key tab while the glue dries is certain to give a correct alignment to the pad. Notice that I do not suggest the use of Pliobond to help glue the pads; they don't need the extra strength.

In order to provide a proper surface for the pads to close upon, you will have to make a simple "counter-bore" in the manner outlined in Figure 64. Since you will be cutting in plastic, the tool can be perfectly well made of soft iron or even brass, as follows. Drill a ¼-inch hole reasonably close to the center of a length of rod that is an inch or so long, as shown in (A). For the sake of clearer explanation, let us then pretend that the next step is to cut it crosswise with a hacksaw as in (B). Careful filing of a set of bevels and the soldering in place of a piece of ¼-inch rod will give you an ordinary flat-bottomed counterbore of the sort drawn at (D). If you were to use this in an electric drill to alter an already drilled ¼-inch hole, you would get a stepped hole of the sort shown in cross section at (E). This gives a flat surface at the bottom of the step, and a pad could indeed be coaxed to seal upon it if you carefully arranged for it to go down into the hole. Such a hole is useless for woodwinds, however, because the pad would have to be lifted very high to open it. In order to get (F), which is the proper kind of seat for a pad, it is necessary to alter the original drilled hole in the cutter with the sort of countersink used to prepare holes for ordinary flathead wood screws, as shown at (C). The

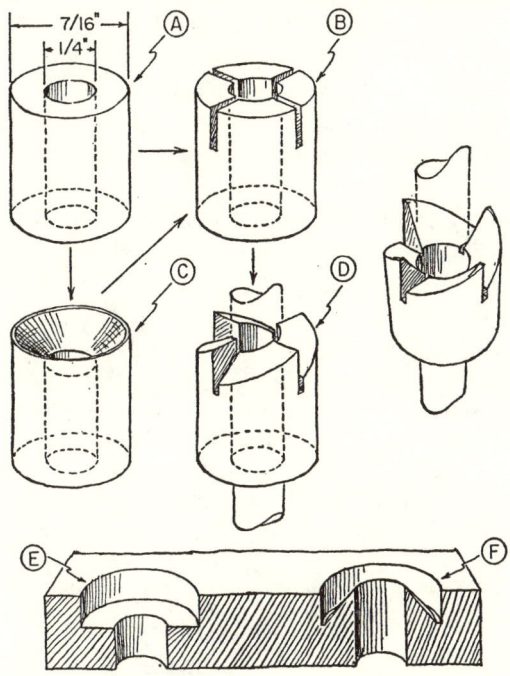

Fig. 64. Method of making a cutter for key pad holes.

saw cuts are put in after countersinking, and then the cutter can be filed as before and soldered onto a rod.

To summarize the procedure: follow sketches (A), (C), (B), and (D) consecutively to make a cutter that produces a pad seat of the sort shown in (F). All the holes to be drilled in our woodwinds will be of the same standard ¼-inch size, so that the counterbore I have just described will serve for any of them that need to be closed by pads. The rod in the middle of the cutter acts as a "pilot" and keeps the cutting edges centered properly with respect to the hole. The object of the cutter is to make a smooth round edge at the rim of the hole; deeper cutting can only weaken the hole and perhaps spoil the tuning of the instrument.

On all woodwinds there are places where one or more fingers have to close an extra hole along with the one under the finger tip, and for this people supply them with "ring keys" or "spectacles," as the French so aptly call them. The clarinet of Figure 67 clearly shows two of these devices. Another way to make them can be seen on the flute: a thin plastic rod is run along next to the right-hand holes in such a way that your fingers press on it as they go down. This is a very poor way to do it, as I found out in 1945 when that flute was made; it took a lot of practice before I was able to press the rod dependably. Apparently our fingers are conditioned *not* to put forces on the rods they find in the key machinery of ordinary woodwinds.

Figure 65 gives a close-up view of the ring-key system that is intended for the three main fingers of your right hand on both clarinets and flutes. A sheet (A) of metal is cut out and marked off with the spacing of the holes it must match before the set of three ⅜-inch finger holes are drilled. To give this rather long item of hardware a little more stiffness, it is a good idea to solder the axle to it on the underside at a

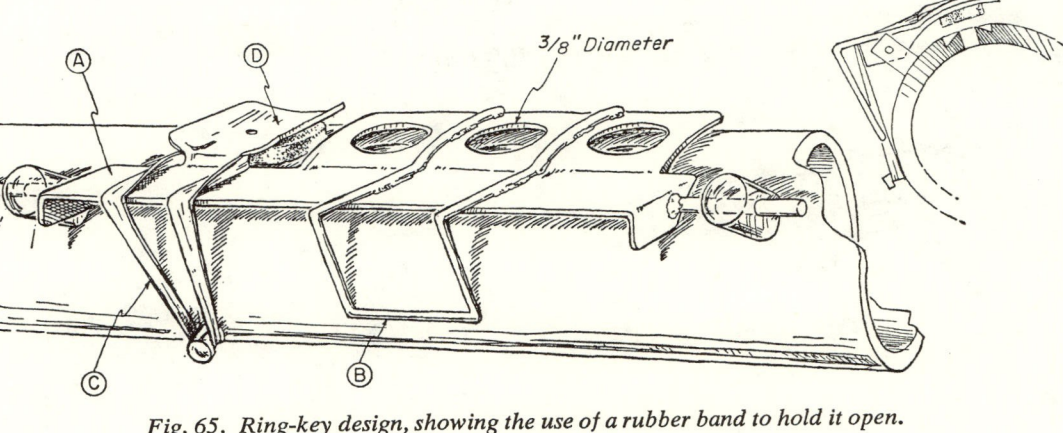

Fig. 65. Ring-key design, showing the use of a rubber band to hold it open.

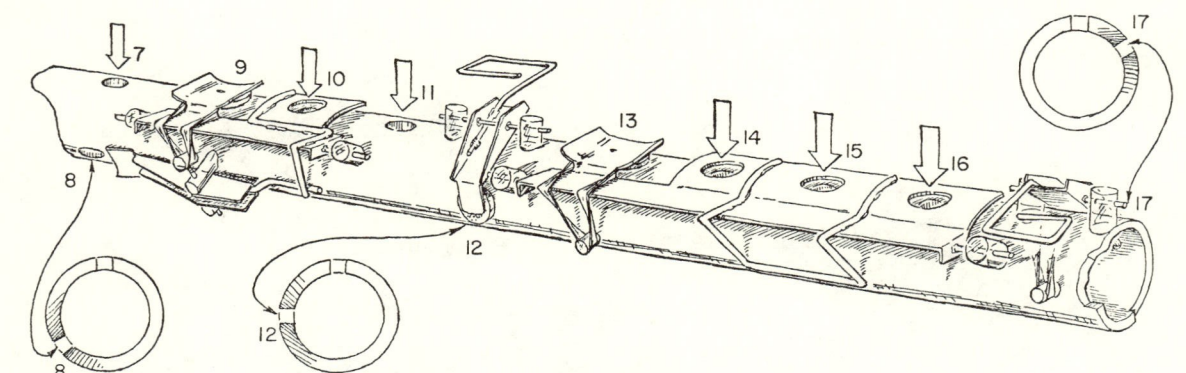

Fig. 66. Over-all layout of machinery for a Boehm-style flute. The large arrows show the finger and thumb holes, while the numbers refer to particular hole positions as explained in the text. The right-hand part of this diagram also applies to the clarinet.

couple of spots, and in addition solder on a U-shaped wire (*B*), which also serves to keep the rings from rising too high when no finger is holding them down. The rubber band (*C*) is put on in the typical way for holding a key open, as I mentioned earlier, and a pad is glued in place under the tab (*D*) after the rings are mounted on the instrument.

Those of you who are familiar with flutes will recognize Figure 66 as a sketch of the way in which the main key machinery can be arranged so as to make an imitation of the Boehm system. The heavy arrows show where the fingers and thumb go, while lower down are three diagrams showing the angles at which the holes are to be drilled for the thumb hole and the holes that are controlled by the G$\sharp$ and D$\sharp$ keys. All the other holes are drilled in a straight line along the top of the bore in line with the embouchure hole, which is to be ⅜ inch in diameter. The number which appears next to each hole refers to something which I shall explain later on. I have not shown the mechanism for the foot joint, as it is seldom used in ordinary playing. If you should want to build it, follow the design for the analogous keys on the clarinet. The C key would be bent from the end of a long axle which turns the lowest pad-covered hole, and the C$\sharp$ key lever is soldered directly to the C$\sharp$ key, which is pivoted on the long axle mentioned.

The lower part of a clarinet is to be made in exactly the same way as the flute, but the upper part is just a little different because of the altered arrangement of the ring-key and thumb-hole leverage. Figure 67 illustrates the mechanical details of the keys and shows the suggested angles at which the holes for the C$\sharp$ and E$\flat$ keys should be drilled relative to the

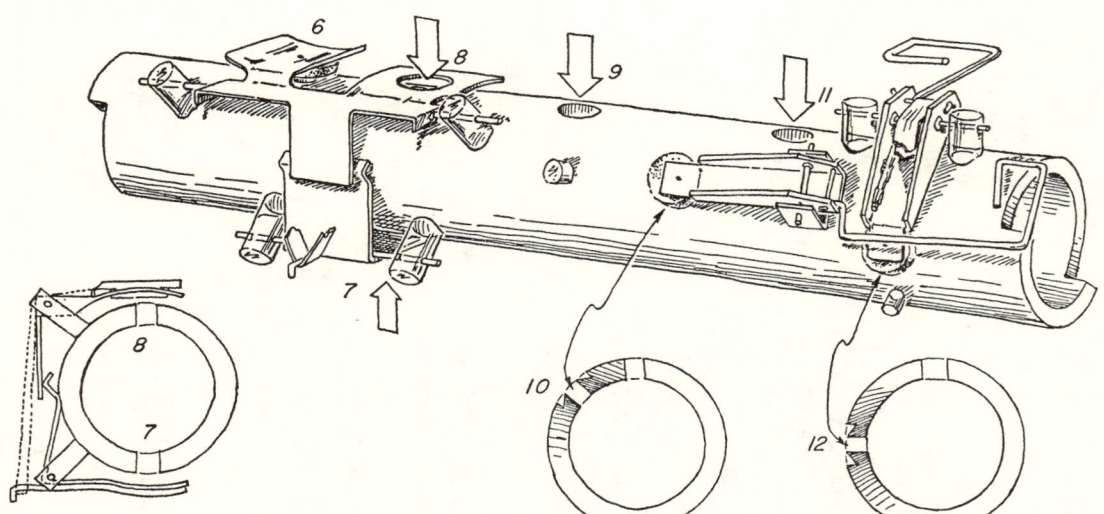

Fig. 67. Upper (left hand) part of the simplified Boehm clarinet mechanism, showing the ring-key arrangement for thumb and forefinger and the mounting of the side key for playing E♭. Again, the large arrows show the finger holes while the numbers refer to the actual positions of the various holes.

main line of finger holes. The thumb hole is drilled directly across from the line of holes instead of at an angle as in the flute. Once again, a set of heavy arrows shows where to put your fingers and left thumb. There is quite a spread between the second and third finger holes on this clarinet because we are doing without an extra ring key which is worked by the second finger. This simplification will affect the accuracy of tuning a bit, but probably not much more than the errors brought in by an amateur drilling job in the first place. For brevity I have not shown how to make the octave key or the G♯ and A keys at the top of the bore, but the method of construction is easily adapted from the traditional way.

The Positions of the Holes

I have computed for you the positions of a series of twenty-five holes, all the same size, which will give a fairly accurate even-tempered chromatic scale if they are drilled into the side of a piece of plastic tubing whose inside and outside diameters are $\frac{9}{16}$ and $1\frac{3}{16}$ inch. The general run of error brought in, among other things, by the uniform hole size is such that the tuning will be more or less usable with tubing which has an inside diameter lying between $\frac{1}{2}$ and $\frac{5}{8}$ inch, provided the wall thickness is not too far from $\frac{1}{8}$ inch. In the table following, these holes are numbered consecutively in descending order of pitch; also given are a letter name of pitch designation, and a nominal frequency taken from the written scale for a piano tuned to A-440. Each hole is labeled with the pitch it will give if it is the highest one of a set of open holes on a flute. Because a flute behaves like a complete conical bore, and a clarinet is more like a cylindrical pipe closed at one end, using these hole positions on a homemade clarinet will give you sounds one octave lower than the pitch labels indicate. Summarizing the general procedure, I shall make specific suggestions on which holes to use.

In laying out the scale on your chosen piece of tubing, carefully scribe a straight guideline down one side over the whole length, and draw a crosswise reference mark on it about 1½ inches from one end. All the hole positions are measured from this spot, which stands for the position of the closing plug on the clarinet-type bore, and for the effective position of the upper open end on the flute. We shall worry about how to arrange the mouthpieces later on. In order to get all the measurements tolerably accurate, it is a good idea to attach your meter stick or steel tape right onto the tubing with masking tape, and to use a sharp steel scriber to transfer the measurements onto the plastic so that they all intersect the guideline. Once this is done, it is a good idea to check your layout with a pair of dividers to make sure that the interhole spacing appears to increase steadily and smoothly from hole to hole as you go down from the reference mark. In order to prevent the lowest note from being out of tune, and to replace the bell, it will be necessary for you to provide several holes below the lowest note you plan to use; these by remaining open make sure the proper sort of "transmission line" is available to the hole system.

Once you have marked the positions of the holes and checked them you are ready to center punch them lightly and drill them through carefully with a small drill whose size is somewhere between $\frac{1}{16}$ and

The lengths were originally calculated in centimeters, as given in the fourth column. The measurements are also given in inches plus decimal sixteenths for convenience in using an ordinary steel tape. For example 5 (13.5) is how $5 + 1\frac{3}{16} + \frac{1}{32} = 5\frac{27}{32}$ is written. You should estimate carefully the decimal parts of the sixteenth.

HOLE NUMBER	PIANO PITCH	FREQUENCY CPS	CM	DISTANCE FROM REFERENCE INCHES + SIXTEENTHS
1	C	1046.50	14.78	5 (13.1)
2	B	987.77	15.73	6 (3.1)
3	A♯	932.33	16.73	6 (9.4)
4	A	880.00	17.79	7 (0.1)
5	G♯	830.61	18.90	7 (7.1)
6	G	783.99	20.02	7 (14.1)
7	F♯	739.99	21.21	8 (5.6)
8	F	698.46	21.50	8 (13.7)
9	E	659.26	23.88	9 (6.4)
10	D♯	622.25	25.36	9 (15.7)
11	D	587.33	26.90	10 (9.4)
12	C♯	554.37	28.51	11 (3.6)
13	C	523.25	30.17	11 (14.1)
14	B	493.88	31.89	12 (8.9)
15	A♯	466.16	33.68	13 (4.2)
16	A	440.00	35.63	14 (0.5)
17	G♯	415.31	37.74	14 (13.7)
18	G	392.00	40.01	15 (12.0)
19	F♯	369.99	42.45	16 (11.5)
20	F	349.23	44.89	17 (10.8)
21	E	329.63	47.53	18 (11.4)
22	D♯	311.13	50.37	19 (13.3)
23	D	293.67	53.41	21 (0.4)
24	C♯	277.18	56.64	22 (4.8)
25	*C	261.63	60.07	23 (10.4)

*Middle C

⅛ inch. Don't forget that certain of the holes are not to be drilled in the main line, but are instead slid round to a new angular position on the tubing! You should now enlarge all the pilot holes to their final ¼-inch diameter with a sharp drill run under light pressure to prevent chipping of the plastic. It is often helpful in drilling in plastic to lubricate the drill with slathers of soapy water or kerosene, and it is imperative that your tools be sharp if they are not to melt the plastic and wad it up into blobs. Recheck all your work and counterbore the three or four holes that will be covered by pads in the completed instrument.

Details on the Flute

Because of the finger stretches required on a full-sized flute, I urge you to build a smaller instrument that is pitched in F so that its written notes sound two and a half tones higher than the piano. The flute shown in Plate VIII is of this sort, and if it were provided with the smoother-working "improved" machinery I have described here, it would make a very pleasant instrument indeed.

The general mechanical details have been explained and diagramed in Figure 66. The numbers appearing by the holes in this drawing are those appropriate to the construction of an F flute. If you want to build it in any other pitch it is necessary only to add 1 to every hole number in order to lower the instrument's nominal pitch by one semitone.

The over-all tube length is 20 inches, and the reference mark is made 1½ inches from one end. Lay out and drill for holes numbered 7 through 20 inclusive, remembering that 8, 12, and 17 are not drilled in the main line of holes. Two of these holes must be counterbored to serve as seats for the pads on the so-called G♯ and D♯ keys. Holes 9 and 13 are drilled in the main line, but must be counterbored for pads mounted on the two "spectacles." Hole 8 is the thumb hole and so is drilled at an angle; it should not be counterbored.

The mouth hole is located on the guide line ½ inch from the reference mark in the direction of the other holes. It also is center punched and pilot drilled before being enlarged to ⅜-inch diameter. Complete the top end of the flute by pushing in a cork plug to a position ⅝ inch *above* the mouth hole. You are now at an exciting and nerve-racking point in the proceedings. It is possible to open and close some of the holes with your fingers, and you can get an idea of the tone quality and tuning of your new instrument! Play around with it a little to make sure all is well before going to the labor of making key machinery.

Your flute when completed should give you a respectable chromatic scale over the range from written D above middle C up to the G or A an octave above the treble staff, with fingerings identical with those of the standard Boehm flute. As a result, you will find a great deal of standard flute and recorder music is perfectly usable and enjoyable to play.

Clarinet Details

It is much more difficult to make a tolerable clarinet than it is a flute because of the greatly increased amount of machinery, and because of the complications brought in by the speaker hole. I shall describe here only the most attenuated sort of instrument, one capable of playing a chromatic scale for one octave beginning at the written G below middle C. If you can coax it up to its second and third vibrational modes, it will play the rest of the clarinet range up from the written D at the top of the treble staff. In other words, the notes between G and D in the middle of the staff are missing. The clarinet of Plate VIII is supplied with these notes as well, and I have already made several comments about them. If your instrument works well, these notes can be added later.

To keep confusion at a minimum, I shall describe a clarinet pitched in C instead of the more usual B♭ and A tunings of the orchestra. You can get these more usual tunings simply by adding 2 or 3 to the hole numbers I give here, since B♭ and A are, respectively, two and three semitones lower than C.

The right-hand part of your simplified clarinet is identical with that of the flute, even to the hole numbers. In the left-hand or upper part of the instrument things are a little different, as shown in Figure 67. Hole 12 is moved out of line and counterbored for a pad as before, and this time hole 10 is moved over and counterbored for the side key pad. The thumb hole 7 is not counterbored, but hole 6 is so prepared for a pad on the ring key which is worked by the left forefinger. I should remark at this point that the rings for left thumb and forefinger normally are kept open by a rubber band that is looped over the ring for hole 8 at one end and hooked to a wire stub soldered to the angle of the thumb ring. This rubber band is shown in part in the main drawing and completely by dotted lines in the detail diagram of this part of the instrument in Figure 67.

You will need a regular clarinet mouthpiece for

your instrument, and a homemade barrel joint to adapt it to your tubing. Let us first get the physics out of the way and then worry about the mechanical details of the joint. Put a piece of masking tape over the flat "lay" of your mouthpiece to seal it and make it watertight. Fill this with water now, and pour the water into a short length of the same kind of tubing used for your clarinet bore. Measure the height of this column of water, and saw off the end of your clarinet at a distance equal to $\frac{3}{16}$ inch more than this height from the main reference mark. This cut is to be between the reference mark and the main row of holes.

When the base of your mouthpiece is butted up solidly against the freshly cut end of the bore and made airtight, you will find that the instrument can be played in reasonably good tune. Figure 68 shows two ways to make the barrel joint you need for this. If you have an old joint that is of no further use to you on a regular instrument, you can easily drill it out to slip over your tubing ready for gluing as indicated in the upper diagram. Failing this, you can take an expansion bit and a piece of hardwood to make your own, in the manner of the lower diagram. Let me suggest that you do this drilling *across* the grain instead of along it, and that the hole be drilled before you trim it down on the outside to a decent proportion. In this way you are less likely to split the wood.

While the procedure I have outlined will give you a reasonably well-tuned clarinet, your instrument will play more easily and give a better scale if the bore cross-sectional area matches that of the mouthpiece. If, for example, you are forced to use tubing whose inside diameter is ⅝ inch, you can make it act almost like the desirable $9/16$-inch tubing by inserting a piece of ¼-inch-diameter plastic rod into it along its

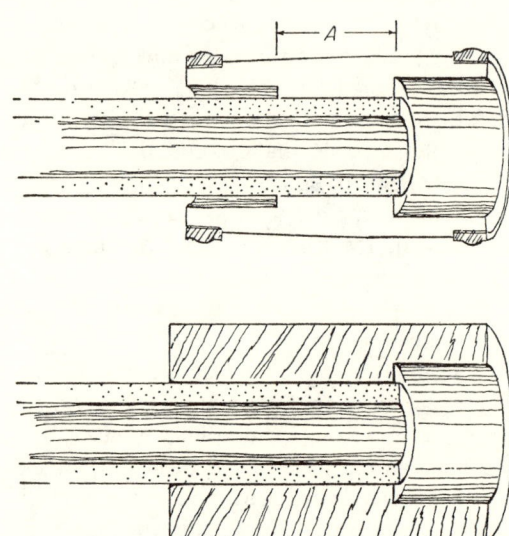

Fig. 68. Two ways to make a connection between a clarinet and a conventional mouthpiece. Top: An ordinary barrel can be drilled out and cemented to the main bore. Bottom: A piece of hardwood can be drilled to suit and glued in place. If you prefer, an ordinary barrel joint can be used, if the clarinet bore is shortened the proper additional amount A.

whole length. The rod "uses up" the right amount of area within the bore, while the strange cross-sectional shape thus produced has no effect whatever (provided, of course, that the rod does not accidentally cover up some of the holes).

If, on the other hand, you are using ½-inch tubing, so that the mouthpiece cavity is too large in diameter for smooth operation, you could line the mouthpiece with a thin curl of plastic sheet, but a much better way is to make a piece of ¼-inch plastic rod about 1½ inches long tapered slightly at one end, and fasten it inside the mouthpiece so that its large end is flush with the end of the mouthpiece joining the bore.

You will have to decide whether or not to use one of these two kinds of rod before the clarinet bore is cut off from the reference mark. The water-filling experiments will have no meaning unless they are done with the rods in place.

The instruments I have been describing will not cause a rush of symphony players to your door. However, you will perhaps learn a little more about, and appreciate a little better, the craftsmanship and perfected judgment which go into even the cheapest of the musical instruments you buy.

Epilogue

Our musical wanderings have brought us back home to our starting place, the interest of a young person in the physical basis of his music. You have met some of the intricacies of music, and some of the ways in which people must steer a course between conflicting requirements. As a matter of fact, almost the whole of this book has been devoted to wading through successive pools of complication toward things which appeared simple at the beginning. We have not been wandering lost in a labyrinth however; the twin threads of musical understanding and basic physics have led us safely, if not easily, past our difficulties.

I hope you will carry away a little feeling for the joys of scientific curiosity, and a realization that a properly asked question is more important to progress than a whole heap of undirected measurement.

FOR FURTHER READING

THE PHYSICAL BASIS OF MUSIC

Hermann von Helmholtz, *The Sensations of Tone,* Dover, 1954
> This book has served as a foundation stone to many studies of musical physics. It is a large book, but, thanks to careful cross-referencing, you can dip into any part of it enjoyably and profitably.

D. C. Miller, *The Science of Musical Sounds,* Macmillan, 1916

Willem A. van Bergeijk, John R. Pierce and Edward E. David, Jr., *Waves and the Ear,* Science Study Series, 1959.

Alexander Wood, *The Physical Basis of Music,* Cambridge University Press, 1913

L. S. Lloyd, *Music and Sound,* Oxford University Press, 1937

E. G. Richardson, *The Acoustics of Orchestral Instruments,* Arnold, 1929

Hector Berlioz, *Treatise on Instrumentation* (Re-edited by Richard Strauss), Kalmus, 1948
> Two composers known for their skillful use of orchestral colors join forces to describe the musical

properties of the various instruments, singly and in groups.

MUSICAL INSTRUMENTS

W. B. White, *Theory and Practice of Pianoforte Building,* Bill, 1906
> A justly respected classic, although the physics lapses from time to time.

A. Dolge, *Pianos and Their Makers,* Covina, 1911

Arthur Loesser, *Men, Women and Pianos,* Simon and Schuster, 1954

G. A. Briggs, *Pianos, Pianists and Sonics,* Wharfdale, 1951
> A well-known loudspeaker engineer comments in a lively and personal fashion on various aspects of piano sound and music.

E. Heron-Allen, *Violin-Making As It Is and Was,* Ward, Lock, 1885
> A quaint, opinionated book, full of misconceptions and errors, but well supplied with fascinating bits of violin lore.

J. W. Giltay, *Bow Instruments, Their Form and Construction,* Reeves, 1923
> In addition to the subject matter indicated by the title, this excellent little book describes simple and ingenious experiments which greatly clarify the physics of violins.

R. Alton, *Violin and Cello Building and Repairing,* Cassell, 1946

Adam Carse, *Musical Wind Instruments,* Macmillan, 1939

FOR FURTHER READING

Anthony Baines, *Woodwind Instruments and Their History,* Norton, 1957

Theobald Boehm, *The Flute and Flute-Playing,* Case, 1922

This book, translated and annotated by D. C. Miller, is Boehm's own account of the development of his revolutionary flute.

Philip Bate, *The Oboe,* Benn, 1956

Geoffrey Rendall, *The Clarinet,* Benn, 1957

These two books are beautifully written monographs on the history, musical properties, and acoustics of their respective instruments. It is a pity that literature of comparable quality is not available in English for the other orchestral instruments.

The Encyclopaedia Britannica is a dependable source of information on musical instruments.

INDEX